Granny @ Work

Granny @ Work
Aging and New Technology on the Job in America

Karen E. Riggs

ROUTLEDGE
NEW YORK AND LONDON

Published in 2004 by
Routledge
29 West 35th Street
New York, NY 10001
www.routledge-ny.com

Published in Great Britain by
Routledge
11 New Fetter Lane
London EC4P 4EE
www.routledge.co.uk

Copyright © 2004 by Taylor & Francis Books, Inc.

Routledge is an imprint of the Taylor & Francis Group.

Printed in the United States of America on acid-free paper.

All rights reserved. No part of this book may be reprinted or reproduced or utilized in any form or by any electronic, mechanical or other means, now known or hereafter invented, including photocopying and recording or in any information storage or retrieval system, without permission in writing from the publisher.

Library of Congress Cataloging-in-Publication Data

Riggs, Karen E.
 Granny @ work : aging and new technology on the job in America / Karen E. Riggs.
   p. cm.
 Includes bibliographical references and index.
  ISBN 0-415-96582-9 (hardback : alk. paper)—ISBN 0-415-96583-7 (pbk. : alk. paper)
   1. Aged—Employment—United States.   2. Information technology.
 I. Title: Granny at work.   II. Title.
 HD6280 .R54   2004
 331.3′98′0973—dc21   2003012529

For Tom Riggs

Contents

Acknowledgments		ix
Chapter 1.	The New, New Deal: The Post-Retirement Era	1
Chapter 2.	Lost Boomers in Space: Aging Workers and the Soft Digital Economy	21
Chapter 3.	An American (Techno) Legend: Women, Age, and the Harley-Davidson Workplace	57
Chapter 4.	"Granny, Go Ahead, You Won't Tear It Up": Central-City Elders Go Computing (with Jacquelyn Vinson and Amy Lauters)	81
Chapter 5.	Use It or Lose It: The Self-Programmable Elder	103
Chapter 6.	Wizards, Space Cowboys, and (of course) Sean Connery: Film Images of Aging Workers	117
Chapter 7.	Who You Callin' Dude?: Magazine Advertisers Discover Older Workers	157
Chapter 8.	How to Win Matures and Influence Boomers: Intergenerational Communication Through Self-Help	179
Chapter 9.	Driving with Dad: Intergenerational Journeys on the Superhighway	207
Chapter 10.	The Digital Divide's Gray Fault Line: Aging Workers, Technology, and Policy	225
Notes		243
Afterword		257
Index		259

Acknowledgments

This book has benefited from the advice of many scholar-friends over the past several years, including (and especially) Mia Consalvo, John Caldwell, and Tasha Oren. It has also benefited from the collegial labors of many graduate students, including Jacquelyn Vinson, Kathy Keltner, Amy Lauters, Jeffrey Smith, Traci Currie, and Joette Rockow, as well as from the help of many work-study students, especially Pat Middendorf. Anonymous reviewers of *Emergences* and reviewers and audience participants at the conferences of Informing Science, Console-ing Passions, Association of Internet Researchers, and International Communication Association's Feminist Scholarship Division provided useful feedback on selections of early work. I am also grateful to the anonymous readers of earlier versions of this manuscript, through Routledge, who helped advance my thinking. Earlier versions of Chapters 1 and 4 were published, respectively, in the collected papers of Informing Science 2002 (Cork, Ireland) and the journal *Emergences*.

As important as the scholarly guidance I have received has been the support of school colleagues, including a "dream team" of office co-workers who have always encouraged me: Paula Carpenter, Jeff Redefer, and Ben Schneider. My dean, Kathy Krendl, a longtime mentor, was tremendously supportive (including the liberal furnishing of high expectations).

In the editorial community, I am grateful to Matt Byrnie at Routledge for believing in this project and encouraging my ambitions to develop it, then providing steady guidance about the writing along the way. The editors and editorial assistants, including Sally M. Scott, Danielle Savin, and Emily Vail, who took care of the manuscript, are also appreciated.

As always, I am grateful to the four living generations of my family, as well as to my late father, all of whom have coached me to do my best and to enjoy it. My 92-year-old grandmother, who still lives on her own, and my 70-year-old mother, who embraces life each day with my wonderful stepfather, are my role models. My husband, Tom Riggs, has supported me extraordinarily. Our children, Austin, Emily, and Jesse, have done their best to be understanding about my frequent absences and are a joy to return home to each night.

Athens, Ohio
May 2003

CHAPTER 1

The New, New Deal: The Post-Retirement Era

> Old age policy is a crime, but worse is the treatment we inflict on the young, to expect to become throw-outs.... Society turns away from the aged worker as though he belonged to another species.
> —Simone de Beauvoir, *The Coming of Age*[1]

This book explores the contours of old age as it meets up with new technologies in contexts of work. It is a rapidly changing intersection, considering that people aged 60-plus spend more time on the Internet than people of any other age group and have shown the largest increase in computer and Internet appliance purchases since 1998. Thirty years ago, such a development would have been unimaginable.[2]

My grandparents left their jobs at the cotton mill in about 1970 at ages 62 and 65. The mill that built their southern river community stood as a red brick symbol of more than forty years of work apiece, and families like theirs walked home from it with uneasy gratitude six days a week. The Chattahoochee River flowed like a moat between Columbus, Georgia, and Phenix City, Alabama. The sturdy mill, hulking over the Georgia riverbank, was everyone's benefactor. Its noisy textile looms welcomed several cohorts of weavers through the century. Their hair was filled with lint from their youth until they collected Social Security benefits. The mill had rescued them from stingy fields of red clay and sand in the country, brought them all survival through the Great Depression, and rendered something that resembled prosperity during the post–World War II boom.

In my grandparents' case, the mill had paid for a five-room bungalow with a shady cement porch where they could sit near the clay road and take in the humid days of retirement. The two of them were exhausted, in fact. My grandfather had a few summers' worth of piddling behind a hand plow in the vegetable garden before dementia shut him down entirely, and my grandmother soon afterward encountered a series of strokes and painful arthritis that have left her relatively immobile today in her early 90s. (Once in a while she rides past the old mill, which is now a conference center and museum.)

Children of my Baby Boom generation could see the writing on the wall back then. For those of us in the working class, it read something like this: *Work hard, retire, fish a little bit, and rest up till your time comes.* Most folks could expect to live only a decade past retirement, if they were lucky, and they could recall many in their peer group who had died well before the gold watch had come. In the late 1980s, my own parents faced this rite of passage by purchasing a thirty-foot motor home, which they would have driven until the wheels wore out had my father's cancer not abruptly canceled their plans.

The act of leaving work in the 1960s, '70s, and '80s was something widely anticipated and thoroughly deserved. It was made especially sweet for folks who could remember their own elders toiling in cotton fields and other undistinguished places until they died. End-of-life leisure was a relatively new phenomenon for Americans, whose life expectancies were creeping upward and who were perhaps the first in their families to look forward to Social Security benefits and private pensions. Similar events were occurring in the countries of Europe, Japan, and other developed nations. The Social Security Act of 1935 established 65 as the official retirement age for Americans, for the purpose of benefit collection. At the time, the life expectancy for Americans was 47 years old.[3] Not until 1964, the last year of the Baby Boom, did the life expectancy for newborn Americans reach 70. Americans who reached "Social Security age," or 65, during that year, could generally not expect to draw benefits.[4] Social Security had, in large part, been created as a Depression-era program to provide an economy-building safety net so that older men could leave manufacturing jobs to make room for younger workers. The gradual reconceptualization of retirement from a brief moment before death to a full-blown period of relaxation and leisure took most of the mid-twentieth century, starting from the Social Security program's beginnings in 1936. By the 1970s, Americans by the millions, largely men, began leaving the workforce earlier, abetted by early-retirement packages and dependable fixed-income investments.

For a time in the 1980s, swelling interest rates on certificates of deposit and lean-budgeted corporations looking to fill their ranks with younger, cheaper labor provided incentives for workers to retire earlier than ever, at least in the middle and upper classes. (Apart from financial inducement, among men the primary reason for retiring early has been poor health. Among women—who usually are younger than their husbands and who retire about three years earlier, accordingly—the main reasons for taking early retirement have been family reasons, desire for increased leisure, and health.[5]) Millions grabbed onto Individual Retirement Accounts with visions of retiring at 55 and taking it easy. Many Baby Boomers smiled at this looming picture on the horizon as they entered their own careers. That was before we learned that the Social Security Trust Fund was on the skids.

The image of the rocking chair—or the Sun Belt condominium—is a fleeting one for many of today's would-be retirees. Fat-salaried men in their 50s who get downsized by big corporations find they still need to work for pay, either because they want to afford their current lifestyle or because they self-identify so strongly as workers. In many cases, their children are hanging on at home (or in graduate school) in their 20s and 30s, and they have taken on other economy-size debts, like the aforementioned motor home. In some cases, those grown children bring responsibilities for grandchildren. Additionally, employers and governments have cut back retirement programs, disability benefit programs, and Social Security benefits. At the same time, older skilled workers are encountering increasing difficulties in finding employment due to disparities between their anticipated wages and demand rates for their skills.[6]

Widowed and divorced older women, along with many older men, dangle just over the poverty line. They work for minimum wage at Wal-mart, long-distance telemarketing firms, McDonald's, and airport garage toll operations. City attorneys retire and pocket their pensions while rejuvenating trial-law careers. Firefighter pensioners collect garbage by day and watch business premises at night. Accountants on early retirement take limited-term jobs as office clerks and consultants to afford the upkeep on their sailboats. Gone is the expectation that 60 means slacking.

Even if you do manage to retire, you find that retirement carries with it flurries of activity that stand in for work. You have to *work* at aging, it seems, if you want to cheat death, taxes, and shame. What used to be pitched as recreation and relaxation is now commodified as that most honorable activity, work. Retirement communities sell the trope of *successful* retirement to buyers, for whom a substantive itinerary translates as a sort of retirement integrity. Not shuffleboard courts and canasta parties at these middle-class sites but lifelong learning with an emphasis on Internet investing, travel, and tai chi. In my dissertation research in a Midwestern retirement community, in fact, I found that elders built a social framework around their television consumption practices, treating their use of "quality" program content, such as C-SPAN and PBS's *Masterpiece Theatre*, as a sort of avocation. These elders scoured the channels for "worthwhile" television that could become a form of capital in their conversations with neighbors as part of an implicit contract to guarantee a successful, worthy retirement for one another. Most of the time, this meant monitoring the information flow from Washington in the most detailed and analytical formats possible. For them, translating news talk into table talk meant democratic participation.[7]

Throughout American culture, of course, work is increasingly defined under the terms of the Information Revolution. Except in the plainest of manual work, and even in certain sectors thereof, it is difficult to find a place where Americans labor without microelectronics. Imagine the changes encountered by today's elders. Workers who today are 70 years old were children at a time when a trip to the moon must have seemed impossible. After all, Charlie Lindbergh's exploits stilled seemed amazing to their parents, many of whom still lived in homes without telephones. This was the day of downtown trips to make layaway installments, the very infancy of revolving credit. Imagine the technological changes seen in the lifetime of such a person, who now may be called upon to retrieve email attachments, swipe debit cards, send faxes, and participate in real-time chat, all in a day's work. They have read about John Glenn going into space not once but twice, the second time as an elder. In many cases, these are people who can remember their mothers churning their own butter.

Aging and Work

Old age is a problematic field, always subject to renegotiation of meaning due to changes in life expectancy; this has never been more true than in the critical first three decades of the twenty-first century, when the proportion of older people is dramatically increasing, with the developed world in the lead. For instance, in Italy, a land of drastically declining fertility rates, by 2025 people over 50 will outnumber people under 50.[8] Leave it to the Baby Boomers to ensure that aging would become sexy, at least from a marketing perspective. It is this cohort that has created a best-seller niche for our elders—Tom Brokaw, Jimmy Carter, and Betty Freidan—to extol the virtues of old age, once a near pariah for popular literature.[9] By January 2002, Amazon.com listed 4,448 responses to queries about books on aging. But, while the onslaught of Boomer elders is the largest contribution to our cultural preoccupation with the subject, that attribution is, of course, not the whole story.

In 1900, the average life expectancy at birth in the United States was 47 years.[10] Individuals we would think of as "elderly" today were indeed rare at that time. Medical advances due to the popularization of germ theory at the century's start and technological progress throughout the century, together with improved diet and safer living and working conditions, made for improved longevity. By the beginning of the current century, Westerners could fully expect to know their grandchildren and perhaps their great-grandchildren.

Popular conceptualizations of "old" in twentieth-century America and throughout the West clearly fell under indictment as older people gradually began to live longer and healthier lives and, abruptly, as the Baby Boomers—in the United States, the 78 million people known as the Me Generation—discovered that they were arriving in what they had heretofore tagged "middle age." Categories underwent revision, and the cultural vocabulary grappled with awkward new names. Euphemisms like "senior citizen" quickly fell out of favor except for use in talking about the most aged groups, such as those over 75. In general, it has become insulting to address a person as "elderly" unless that elder resides in a nursing home and presumably suffers from dementia and other forms of incapacitation. "Older adults," "mature adults," and simply no tag at all have become preferred, and even these choices are controversial, especially for people with money. Middle-class people in their 60s don their Reeboks and listen to their Rolling Stones CDs and don't want to self-identify as seniors unless it means exercising their AARP discounts on Medicare supplements or their lodging at the Holiday Inn. The change wrought by the Baby Boomers' arrival into their 50s was exemplified by the American Association of Retired Persons' introduction of a second magazine for its younger membership, who did not care to have *Modern Maturity* arriving in their mailboxes. The magazine? *My Generation,* its title taken from a '60s hit by The Who. Acceptance of aging on these special, even "cool" terms was further legitimated by the magazine's first cover photo and profile subject—the "cute" Beatle, Paul McCartney, age 57. For Boomers, who may have internists, colon screenings, retirement portfolios, and grandchildren, getting older does not mean being old. After all, their own kids are still going to concerts to see Mick Jagger and Tina Turner.

Nicknames for "old" carry distinct burdens for women and men. Women know that youthful attractiveness is sheer capital that has spilled out of their bodies. Because many women among the generations who are now of advanced age have little other capital at their control, they shrink from age-related labels. And because women are four times more likely than men both to live alone in their 70s and 80s and to ultimately be institutionalized, "old" is a label they cannot afford if they wish to preserve their autonomy against those who threaten it, such as their own children.[11]

Old age in twentieth century developed-world societies has been increasingly perceived as a time of retirement, not a time of paid work. What recently has been considered late middle age, between 50 and 65, has been perceived socially as a time of steady decline in productivity, a combination of the mind/body slowing down and preparation for retirement. But data from manager surveys, physical productivity studies, and other sources counter this stereotype, demonstrating that productivity falls only marginally for most people in this age group.[12] What does change is cost to employers for health plan participation, rising from about 5 percent on average for all workers in the United States to 30 percent for older workers, and escalating.[13] Workers are simply less affordable as they age, and they perceive their fringe benefits increasingly as golden handcuffs. After all, it is unlikely that a 60-year-old man with a history of heart disease or cancer, for example, will be able to find affordable replacement insurance if he loses his job, no matter how satisfying or unsatisfying he might find the work.

Generally, economic forecasters cheerily spread the news that today's elders are more affluent and better educated than any previous such group, but that is misleading. Statistically speaking, Americans age 65 and over do hold greater wealth than younger citizens, a phenomenon that has similarly occurred in other developed nations.[14] But, as Richard Disney points out, publicity of such affluence has concealed a tremendous growth in economic inequality that is spreading among the elderly. In Britain, for example, the median income in the top quintile of pensioner income distribution is four times greater than that of the bottom quintile.[15] In another example, a study of workers in the United States, Thailand, and Taiwan in the 1990s, researchers found that income inequality increases among ethnic groups as their cohorts age.[16]

It is logical to conclude that the combination of increased life expectancy and changes in the workforce makeup of developed-world societies warrant the construction of new models for workforce and pension benefit participation, with emphasis on flexibility across age segments. Harry R. Moody, a proponent of the abundance view of aging, has called for social policies that retrain and create jobs for aging workers who have lost out in the postindustrial economy. (Although Moody's contribution came in 1988, it is only made more significant by the escalation of emphasis on technology in the workplace over the last decade and a half.) Moody's lifespan development model encourages a break from inflexible linear life and career planning. As Moody observes, most people are able to work for several years past 65, although they generally desire flexible work calendars for a variety of reasons, including a wish to pursue travel, hobby, and family opportunities and the need to accommodate health demands. Similarly, younger people, who currently are pursuing full-time work to meet the demands of income need and career track expectations, could benefit from segments of time outside the workforce, without penalty, for the purpose of raising children or caring for an elderly relative, for example.[17]

New models that would creatively reconstruct workforce participation likely would result in employers investing more in training older workers in high-value contemporary skills (i.e., technology-driven work) and democratizing such work across age segments. In other words, instead of seeing a "natural" association between newly minted younger workers and high-reward, high-tech jobs (versus a similar association between "un-retrainable" older workers and low-reward, low-tech jobs), employers will hold a more creative palette of role design. It will contribute toward clobbering the myth of "aging in the country of the young," as Moody puts it, but only if fairly implemented to counter middle-class careerism that keeps disadvantaged workers down.[18]

Revised models of work opportunity could contribute significantly to stemming both the feminization of poverty, generally, and the feminization of elderly poverty, particularly. It is widely known that women increasingly head households in developed countries, make less money than men, and are more likely to live in poverty (with children). What is less widely known is that older women, who are more likely to be single than younger women, are especially vulnerable. Women 65 and over constitute only 60 percent of the elderly population but make up 75 percent of the elderly poor in the United States, for example. They spend one-third more on housing (40 percent of their income) than single elderly men. After they pay for health care (spending twice as much as non-elders), little is left.[19] If workforce systems encouraged the participation of women over 65 as long as their health permitted them to work, such poverty would be alleviated. Even if they reach the status of the so-called oldest old, 85-plus, they would have enjoyed the benefit of extended career participation and would be less likely to land in poverty.

Robert Butler, a celebrated geriatrician, coined the term "ageism," defined as a conceptualization of age in chronological years and based on beliefs that old people are ugly, sickly, and unproductive.[20] Ageism becomes age discrimination when these ideologically based beliefs about chronological age are used to systematically deprive people from opportunities and resources that others enjoy, including jobs.[21] Age discrimination is proscribed against by various national policies, including, in the United States, the Age Discrimination in Employment Act, which prohibits mandatory retirement except in certain occupations, and, in Canada, the Charter of Rights and Freedoms. Canadian researchers Julie McMullin and Victor Marshall have established that the experience of age discrimination in employment practices in Canada, similarly to the United States, is complicated by social relations such as age, class, gender, and race/ethnicity. In their study of Montreal garment workers, they concluded that older workers are quite aware of ageist practices that overlook them or otherwise marginalize them in the workplace. For instance, lower-class workers in marginal positions are further marginalized by advancing age, as are women and minorities. They might be overlooked for advancement opportunities, overtime, and other privileges. Furthermore, because of their positioning in particular cultural contexts, the workers feel diverse degrees of disempowerment. Workers who occupy especially marginal positions might feel there is little or nothing they can do to counter discrimination, and they tend to feel resigned to its effects. Such workers feel powerless in their dealings with management and their ability to play productive roles.[22] Such incidents underscore the complications that older workers encounter in their bids to achieve and maintain vitality among colleagues and competitors in the world of work.

Butler has cited a positive correlation between productivity and health, and, conversely, between unproductive living and illness. "Unless we begin to perceive older persons as productive, their lives will be at risk," he notes. "They will be seen as a burden."[23] Aging, health, and productivity are interlinked with socioeconomic status and are especially influenced by occupation, education, national origin and residence, race, and ethnicity, Butler continues. He concludes that these issues are not to be solved uni-generationally but that the generations must work together to ensure opportunities for productivity among the able-bodied aged.[24]

"Productive aging" is not without its critics as a concept, however, and for good reason. The concept tends to convey the expectation that if we all practice productivity in our later years—successful aging—then problems such as crippling arthritis, dementia, depression, loneliness, and, perhaps most especially, prejudice against the elderly will vanish. Productivity cannot monolithically occur, because some people are unable to practice it. On another front, the notion of successful or productive aging perpetuates a market logic, underscoring the notion that every citizen's duty is to perform for the economy, as Caroll L. Estes and others have observed.[25] Such logic promotes inequality among the elderly, because, as Butler himself has noted, education levels and race/ethnicity influence opportunities for obtaining the plum work that will allow us to age successfully.

As a society, we must learn to honor elders not simply for their role in decreasing their own economic burden through more work but for the complexity of roles they play and have played among us. While it is heartening to see people over 65 enjoying workforce participation in a strange, new technological environment, it is just as significant not to spurn those who are unable to lead such lives—and to ensure that they are not severed from participation in society as a result. Butler and proponents of productive aging argue for intergenerational alliances to encourage elders to contribute to the maintenance of society. Estes and similar political economists who critique the productive aging view argue for citizenship models that promote society's interdependence rather than liberal individualism.

Taken together, these approaches can teach a valuable lesson: Advanced age is a complex dynamic, fraught with both opportunity and peril. It is important, but insufficient, to practice policies designed to eliminate simple ageism. The range of differences that create both our wonderful distinctions from one another and our differential access to opportunity play significant roles in the aging process. Policies aimed at enhancing economic opportunities for aging workers and retirees do damage to some if they are written in isolation. Policies designed to level economic access across lines of gender, race, and other cultural differences—throughout the generations—will enhance successful aging for everyone. For example, the United States could take a cue from Sweden in formulating its retirement benefits policies. Instead of offering pensions and health benefits on the basis of work history, as the United States does with Social Security and Medicare, Sweden does so on the basis of citizenship. Economic differences that are due to gender, race, and sexual orientation are assuaged by such a policy. Chapter 10 will return to these matters and will suggest an agenda for intergenerational success.

New Technologies

Cultural scholars have been increasingly occupied with understanding the role of new electronic media technologies in the everyday lives of developed-world citizens. Among their important works have been attempts to understand and explain the paradoxically empowering and subjugating role that such technologies play. The most successful of this scholarship acknowledges that computers and related technologies—cell phones, personal digital assistants, and the like—tend to deepen structural inequalities brought about by the systems in which they are implemented, most notably late liberal capitalism. At the same time, this line of scholarship acknowledges, these technologies contain inherent possibilities for at least partial equalizing measures of agency. Even as computers are used to dehumanize and deskill workers through what Chris Carlsson has identified as a "creeping monoculture"[26] (think Bill Gates), it doesn't have to end this way. As Pippa Norris observes, politics as usual may be altered by digital technologies through shifting the balance of resources among political interests, reducing costs of gathering and disseminating information, and benefiting fringe activists in the process. These measures, Norris argues, can shift the balance of resources from the holders of land and capital to parties with skills and expertise.[27]

In the case of the aged, the picture is mixed. Because most skills and expertise associated with computer technologies and the Internet are acquired through schooling and other activities most frequently associated with younger generations, people over 55 still generally lack such assets.

Contemplating the Digital Divide and the "information poverty" that is associated with it, Norris likens the Internet to gunpowder. In this compelling metaphor, she observes that gunpowder deepened the power differential between societies that had and did not have the technology until it became broadly diffused. The question, though, for Norris, is how the Internet will become diffused in various societies around the world. She speculates that, while less-equipped social classes, age groups, and ethnic minorities will enjoy some degree of eventual "catch-up," the Digital Divide will never be completely erased because structural relations will prevent a complete leveling out. Some people will simply never have access to this asset for diverse reasons, some economic, some social, some cultural.[28]

But the horizon is not all doom and gloom. Elders are busy on the Internet. Beyond researching their genealogy, emailing grandchildren and snowbird pen pals, and looking for love online, elders are working (which does not preclude the aforementioned activities as acts of work). They comparison shop for cruises, second cars, and sweaters, and they research stock activity as acts of intelligent consumption. They rely on telemedicine to become increasingly proactive patients and Medicare consumers. They engage in lifelong learning to sharpen their skills and intellect. And, increasingly, they use the Internet to learn about technology itself.[29]

Complicating Age

If the intersection of age-work-technology wasn't sufficiently complex to understand, we must figure that it is complicated by other cultural issues. Gender, sexuality/sexual orientation, race/ethnicity, social class, national identity, and religion also mark this experience. Scholars concerned with any or all of these categories are curious about how people occupying particular cultural positions find themselves embedded in the historical practices and developing changes associated with work and new technologies.[30]

Thinking through issues involving older workers gets help from feminist approaches, which appreciate the tenuous but almost predictable nature of women's "place" in the work-technology contest. For example, in her book *Shaping Women's Work,* Juliet Webster demonstrates problems that flow from women's absence from the creative process; it is here that technologies later to be used in women's work are first developed and implemented.[31]

In her own critique of technology, Cynthia Cockburn has demonstrated similar points about the distance of women from locus of design and the resultant black-boxing of instruments of women's technological labor, from microwave ovens to assembly-line machinery. Women's work in both domestic and paid contexts gets regulated by common-sense structures that snub women.[32]

Can we similarly struggle to understand the unfolding relationship of older people to new technologies? How do these technologies enter into the world of work for older adults and how do they influence (or get taken up in attempts to influence) the status of older adults? And, perhaps most urgent: Given the rapid turn to both a technology-driven economy and an aging society, how can we understand the many points at which these two speeding trajectories will cross over one another, now and in years to come? How will rapidly updating technologies of work get along with an older body of workers?

We can approach these questions from a number of fronts. First, it is useful to ask how ageism might also come into play. If feminist scholars mourn the distance of women from the places where technologies are developed and implemented, what does this mean for research about older adults? As reported in the U.S. Department of Commerce white paper *The Digital Work Force: Building Infotech Skills at the Speed of Innovation,* three-fourths of computer systems analysts and computer scientists, and four-fifths of computer programmers, are under the age of 45. The study's authors found that almost half of information technology (IT) managers in their 20s and 30s—accounting for many of the sector's gatekeepers—had never hired a worker who was over 40.[33] Granted, the role of education accounts for much, but not all.

Bias against older workers is so firmly entrenched in the United States that legislation had to be implemented to warn against it in the Older Americans Act of 1965. It is useful to wonder what employers, supervisors, trainers, software engineers, and other gatekeepers of work and new technologies are thinking about the roles that older adults might or might not play in their worlds. In these spheres of increasing influence, it is important to know something about the structures in which older adults have labored previously and how changing work conditions likely will either keep such structures intact or lead toward change. Not to be discounted is simply the demographic shift, wherein the proportion of older adults is mounting furiously. Some countries, in fact, may exceed a median age of 55 by the year 2035.[34] People are living longer after retirement, and fewer new employees are being turned out to take their places. Many employers throughout the developed world have begun to see older workers as a compliant source of cheap temporary and part-time labor in flush times, which is both enabling and disempowering to elders.[35] For example, in Japan, men are likely to be forced to retire at age 55 and must find other work, often temporary and almost always at much lower pay.[36]

This book attempts to provide a context in which scholars, activists, and others might begin talking about the changing role of work for older adults in a high-tech economy. Instead of offering a statistical breakdown that can be generalized to our entire older adult population, it tells the stories of real people associated with this complex set of concerns, demonstrating how difficult it is to paint any definitive sort of portrait of aging in American culture. Its primary usefulness might be in the recognition it offers that, like the rest of us who are reeling from the velocity at which change is arriving in contemporary life, elders are facing multiple tensions, consequences, and challenges and are meeting these with varying outcomes.

I am especially interested in the stories of the elders whose high-tech adventures are not celebrated in the pages of the *Wall Street Journal* and not championed on the home page of Seniornet.com. The middle and upper classes of retirees whose retirees have discovered the World Wide Web as a hobby destination have certainly been able to enrich their lives, a wonderful and laudable accomplishment. But the stress points of elderhood are much more provocative for me: the older adults who "survive" in the computerized pink-collar shop. The working-class unskilled laborer who craves computing classes for another, better chance at paid work, or the one who shuns computing from her home because of its forbidden and frightening threats to personal safety and security. The Gen X office manager who must bring older adult workers online or turn them out to pasture.

Most generally, these kinds of stories are complicated by struggles associated with other arenas of difference, including class, gender, race, and ethnicity. These struggles, when combined with the challenges of advancing age, can set the scene for serious negative consequences. But failure is not destiny. People caught in the "triple bind" of gender, race, and age disadvantages in the work place can thrive. On the other hand, the much-maligned white male worker, predictably in his peak earning years between 50 and retirement, can be awfully vulnerable to harsh employment conditions in which information technologies' roles bear on the deliberations by those who will decide his fate.

bell hooks has written about the uphill climb in privileging class as the concern of academic writing. She is right, of course. It is difficult for middle-class feminists to engage in ethnographic research, or research from other methodological traditions, without making references to people they understand well—other middle-class women. hooks insists that "class matters" as much as race and gender, especially in light of the ever-gaping digital divide.

I have the same feeling about age. In the conference papers I hear and the essays and books I read, feminist researchers of contemporary culture most often refer to themselves in the third person somehow, and very few of them are currently over 50. Thirty-something feminists write about the Oxygen channel and the pleasures of shopping. I have heard more *Buffy the Vampire Slayer* papers written by young women in graduate school than I can keep straight. I have heard the youngest students at communication conferences deliver papers on teenage "girl power."

Such distinction makes complete sense, and it's a phenomenon to celebrate. Pieced together, as the quilt of cultural research so beautifully invites its readers to do, these writings from variously aged researchers helps to construct an intergenerational understanding of women, of feminism, that enriches a broad conversation. But, on the whole, the conversation says little about the role that older adults—women and men—play in contemporary culture, both in its production and in its consumption.

I believe a chief reason for this short shrift is that the idea of developing theoretical discourse on subjectivity just becomes unwieldy at a point. If we dare to cover, in our broad discussions, the entire package of the "holy trinity" of gender, race, and class (perhaps substituting sexuality for one of these, probably class), it becomes too onerous to think about other differences. After all, our conference papers must be limited to twenty-five pages. Studies focusing on young people, at least, can focus on children, teens, and young adults in their role as emerging consumers.[37]

Other reasons exist for the near absence of older age from the broad discussions of gender, race, and class. Scholars, like other members of our particular culture, easily identify as first gendered or as members of a particular race or ethnicity and, at least in cultural studies, the fact that they might occupy marginal positions because of such identities propels them into study and writing. As hooks has observed, this has been less apparent along the axis of social class. In Western society, people—even academics—are unlikely to identify first and foremost as old, middle-aged, or young. Age as subjectivity, unless it calls attention to itself through reminders of the limitations, is simply not salient for most people most of the time.[38]

Especially in North America and Western Europe, where the category of middle age has expanded terrifically to suit the sensibilities of sexagenarians whose parents are living longer than expected into old age, elderhood as an experiential concept is increasingly rejected. In countries where youth is supreme, growing numbers of people who would have, two decades ago, been considered elders at their present age now think of themselves as relatively young. This reprieve provides no small solace to a generation careening into the latter segments of life.

Old age, as it embodies the perceived loss of vitality and the decay that signals death to come, holds no allure for most scholars. As Margaret Clark observed about her own field more than thirty years ago, studying the aged body is anathema to the anthropologist because it is "somewhat akin to necrophilia."[39] We are grossed out by the decay that greets us. We are reminded that it will happen to us.

Some wonderful scholarship from the humanities has taken place in the field of aging, however. Some of the most important contributions, many of which have influenced my own thinking, have taken place in connection with the Center for Twentieth Century Studies (now known as the Center for Twenty-first Century Studies), located at the University of Wisconsin–Milwaukee. Former Director Kathleen Woodward's work on the images of aging, especially in literature, has been at the center of this project on age. One of the most compelling points that Woodward and many of her colleagues press us to see is that we cannot study late life without considering its roles relative to the occupants of other stages.[40] Intergenerational relationships are part of everyday life. We cannot theorize Grandma without also theorizing the many-aged people who are her kin, the providers who meet her needs and desires, and those who might exploit her. (I am not necessarily suggesting that these categories of people are mutually exclusive.)

The argument to study age, not just aging, is especially powerful in light of the changes being wrought by longevity, lower birth rates, and the aging of the Baby Boomers. There simply are going to be, in many countries around the globe, many more elders than we are used to having. Their numbers are uncomfortably rubbing up against those of younger workers, who labor to pay their pensions; care givers, many of whom belong to different class and ethnic groups; and children, whose very definition is being reimagined because of their changing ratio to elders. Never before at any time in the United States have there been so few children compared to so many elders.[41] These younger groups of people and elder generations are in the midst of many struggles to invent new intergenerational relationships that work, and the models are often hard to locate in practice. For example, in her work on care giving, Sharon Keigher has found that, increasingly, elders in need of (and able to afford) services are suburban whites, and paid caregivers are younger people of color living in urban areas. This presents much more than a transportation problem; it is also a larger one of deepening cultural and social divisions.[42]

Nowhere is this discomfort in inventing new intergenerational relationship models more pronounced than in the age-work-technology nexus. Elder workers often feel pressed to step aside in the workplace to make room for freshly trained, computer-hip personnel to spread their wings. A sliver of a generation of young workers is coming to resent having to "prop up" an unprecedented swell of Social Security pensioners.[43] In greater numbers than ever, 50-plus citizens return to complete undergraduate degrees or seek graduate education. Seated alongside pimple-faced "traditional" students, they turn the classroom dynamic on its head.

For many elders, notions of work are entirely bound by the concept of place; for their juniors, it is not. On the other hand, for the juniors, the Internet clearly is a place; for their elders, it is often an opponent. With all the rules changing faster than we can keep track, high-tech work in this postmodern era demands intergenerational bridges for which no blueprints have been drawn. The old rules were long ago thrown out (gold watch, respect your elders, honor seniority), and the rules for Generation X are that there are no easily discernible rules (dot-com millionaires, dot-com layoffs, downsizing). The "rules" for Generation Y, the generation following X, are not yet entirely discernible.[44] Part of the mission of this book is to help draw some of those new blueprints for intergenerational bridges.

One way of working toward formulating bridges is by inviting phenomenology and disability studies to help broaden our understanding of elders' encounters of technology in the workplace. Phenomenology is concerned with reality as lived experience, and many practitioners of disability theory enlist this approach because they are interested in how people with disabilities live through bodies that are always a site of struggle and always subject to definition by others. As the phenomenologist Drew Leder notes, "Insofar as the body tends to disappear when functioning without problems, it often seizes our attention most strongly at times of dysfunction."[45] Likewise, following phenomenologist Maurice Merleau-Ponty, disability scholars Thuy-Phuong Do and Patricia Geist have noted, "Society encourages those with undamaged bodies to speak."[46]

One of the ways in which "society" encourages those with undamaged bodies to speak is through the implementation of technological innovation. Interfaces for accommodation of disability have been added on to computer technologies, to be sure, but they do not promote facile use in most workplaces. Arthritic fingers struggle to keep up with those of the acrobatic Xers. Bifocaled eyes squint at screens that are too dim and low-resolution to read in one's 50s. Youth can speed down the Information Superhighway. This is one highway on which *lack* of speed kills.

Disability is framed by its constraint on one's experience of the world. Many of the people who talked with me as I prepared to write this book spoke of their encounters with computers in the workplace as occasions in which their disobedient bodies—crimped by advancing age—restricted their ability to enjoy the freedom that younger colleagues and relatives could exercise. Whereas opening a sophisticated application, such as an Internet browser or PowerPoint, seemed to be akin to opening the throttle on a motorcycle for these young folks, it could be a stultifying, even painful one for less experienced, slower-reacting, less confident, sensory-impaired elders. "Humiliating," "emasculating," and "public loss of my competitive edge" were some of the descriptors I heard about this event. One woman in her 60s told me that performing such operations in front of more adept young colleagues made her feel as if she were going to fall. "Fail," she corrected herself. A Freudian slip, I wondered. All too commonly, injuries from falls precipitate elders' withdrawal from public life and shunt them toward a lifestyle of inactivity. Fear of falling and fear of failing are often, for many elders, essentially the same feeling.

It is this awkwardness toward computers that defines many older people's response to technological change, but, again, this phenomenon is not to be overgeneralized. Many of the women I talked with told me that the moment their offices moved online and incorporated word-processing programs, it was a liberating experience. It allowed them to spring up the career ladder in the 1970s and continue to enjoy the fruits of those successes today. Growing up in a sexist society, they had been trained as typists in high school, unlike their male counterparts, and, fortunately for them, they did not suffer from arthritis in their fingers during their 50s and 60s, as so many women do. On the other hand, many of the men I talked with in their 50s and beyond felt awkwardness toward computers. They might ably operate the mouse, but they had no idea how to type and were so either left to "hunt and peck" or enlist the support of a secretary ("office wife") or wife at home.[47] For these men, encounters with the computer made them feel slow, ignorant, irrelevant—old. The luckier ones among them had forged relationships with support staff that made them champions of the machinery. They brought slide shows to meetings that someone younger and female had produced, for example, and they received congratulations for yielding research results from the Internet with unproblematic alacrity. For men in this age group, especially, gender has its benefits, and one of these means that you are more likely to be the boss.

But having been the boss for some time, even for a man otherwise well qualified, does not ensure continued success in a changing technological war zone, where younger men in their early 30s come into the shop spouting the vernacular of the Internet. As several of the men who spoke with me complained, they were not simply replaced but their jobs rendered "obsolete" by "young Turks" who were brought in to make their operation more relevant to a company that had begun not to resemble itself. As one man said to me: "The speed was dumbfounding." He meant the speed at which the "Turk" could operate his machine, and the speed at which the gray-haired man subsequently was on the street. "I'm used to being more careful, more purposeful in my work," the man later said. "What I took to be care in my work got me accused of behaving like a senior citizen. Moving too slow."

Accommodating disability theory and a phenomenological approach to consider elders' encounters with computers at work is not meant to underscore stereotypes of the aged as decrepit. It is useful to point out, however, that a computer-equipped lifestyle is readily accessible to people who are more comfortable with quickness, multitasking, and instability. Research continues to demonstrate that advancing age brings a preference for just the opposite: deliberateness, purposefulness, dependable routine. That the machines we find in office spaces have been designed by young people with young people's bodies and senses in mind only complicates the encounter for older workers. It is logical to see why that, the older workers are, the more likely they might be to feel like fish out of water.

This book takes up the experience of aging, not simply as elderhood or old age. I am interested in what it is like to feel that one is aging in the workplace and how one might experience older age relatively. For example, a popular stereotype of Silicon Valley is that it is populated by young, white men in their 20s. I attended a talk by a man in his early 50s who works in one of these stereotypical software shops. He reported being twice as old as any other employee—older than most of their mothers!—and he felt positively ancient. I met another woman who is the same age as this man, and her circumstances are completely different. She works as a nursing assistant in a residential home for the elderly. At the end of her shift, she reports, she cannot wait to get out, among "young" people. "I'm so tired of old people, I don't know what to do," she remarked.[48]

I present the stories and images of people who are in their 50s and beyond, not to muddy the waters about old age's complexity but to underscore the point that, as the longevity revolution lengthens life, there are greater differences among elders themselves to explore. As I've already noted, people in their 50s, and most people in their 60s, at least in developed-world societies, are unlikely to think of themselves as old (even if that's what many Generation Xers and most Generation Yers think of them). It was about three decades ago that sociologist Matilda W. Riley established parameters around the categories of "young-old" (ages 65–74), "old-old" (75–84), and "oldest-old" (85 and over).[49] But just as the cut-points between generations are being reallocated by demographers, new issues about work and technology are being raised for discussion.

For example, the Older Americans Act protects persons 40 and over. Implemented in 1965 for a minority of workers, it soon will protect the majority from discrimination. What are the ramifications of leaving the act unchanged, or of changing it? As another example, 75 percent of occupational injuries from repetitive stress involving days away from work in the United States occurred in workers 44 and younger.[50] Most of these are injuries among nonmanagerial workers, mostly women, and many result in permanent disability claims. What sorts of practices and policies might result in better product design or other solutions, and what effect might these ultimately have on economies and organizational structures? Finally, can the outdated Older Americans Act be revised to remedy such situations as this? I will suggest answers to these and related questions through this book in hopes of enlivening the dialogue about aging, work, and technology that policy makers, academics, and activists are joining.

Content of the Book

Chapters 2, 3, 4, and 5 bring together stories about the experiences of aging workers in the digital economy.

Chapter 2 looks at several popular Web sites in the United States and several other nations that focus on the intersection of late life and work. The chapter draws conclusions about how older adults are depicted and addressed through the assembled discourse within these sites. In it, I include results of email dialogue with some of the elders who use the interactive sites, focusing on how they frame their own experiences, participation, and plans.

Chapter 3 looks at the place of microcomputing in a typical corporation as it is encountered by diverse women at mid-life and in the later years of work. In its depiction of older women's work with new technologies at Harley-Davidson in Milwaukee, Wisconsin, the chapter provides confirmation for notions about the embeddedness of technological change in webs of complex social relations, where class relations are challenging to understand.

Chapter 4 probes ways in which older, central-city women have encountered computing, both through desires for improved working conditions and through an interest in self-fulfillment and preservation of youthfulness. The chapter illuminates conditions that lead some working-class elders to see themselves through the struggle of acquiring access to new technologies even as they lead others to back away from what they see as a radical arena.

Chapter 5 examines the comments of people over 55 who answered my Internet query for stories about transitional moments with new technologies in the workplace. This chapter highlights the special vulnerability and creativity of aging people in the digital economy.

Chapters 6, 7, and 8 consider three popular discourses on elders, work, and new technology. Chapter 6 offers a sampling of popular Hollywood film treatments of the subject and, in doing so, makes some observations about how other avenues, such as gender, race, and social class, intersect with this central concern. Looking at eighteen Hollywood films that were box office successes between 1950 and 2002, the chapter identifies three thematic commentaries on aging, work, and technology: revisionist history, the fairy tale, and the son becoming the father. Films as diverse as *Cocoon*, *Death Becomes Her*, and *Space Cowboys* constitute the sample from which this textual analysis is drawn.

The chapter concludes with a brief annotated listing of other films in which these three elements are also prominent and identifies other thematic commentaries in film using these concepts.

Chapter 7 analyzes the depiction of elders in advertising in a variety of magazines directed at audiences both including and excluding elders around the nexus of information technologies, work, and older adults. Once entirely absent from advertising in which computers and related goods and services are marketed, images of elders are now plentiful. The implication is sometimes one of empowerment, sometimes not.

Chapter 8 looks at relevant themes in the growing body of self-help literature that is available to consumers and decision makers. From books geared toward "dummies" and "idiots" to best-sellers from influential management gurus, millions of published volumes now help gatekeepers and job seekers alike grapple issues surrounding the aging worker. Within this body of popular literature, we can see clear modes of explanation of the Boomer-led longevity revolution as a "scarcity problem" or an "abundance opportunity." The chapter concludes with a comparison of the two book styles' addresses of their readers: the first as aging subjects seeking some form of cultural power, the second as decision makers studying how to maximize their power over elders. These two discursive strategies are seemingly at odds but often reinforce each other.

Chapter 9 draws thoughts on how this intersection of work, new technologies, and advancing age can become less hazardous to navigate. It organizes the policy-related questions raised by the earlier chapters and offers answers to them. It includes advice from thirteen pairs of elders (55 and over) and twenty-something year-old undergraduate students who participated in a semester-long study about Internet use, aging, and work in "intergenerational teams" (a parent or grandparent of each student).

Chapter 10 concludes the book with how lessons from this intergenerational project can contribute to agendas within the spheres of policy, education, employment, product development, and activism.

CHAPTER 2
Lost Boomers in Space: Aging Workers and the Soft Digital Economy

Member Information
Phil
Instantly receives group messages in inbox.
Role: member
vintage diarist@hotmail.com
Hi! I'm 58. Worked 25 yrs. in public sector. Retired 12/01. Wife retired too. Thought we had enough. However, familiar story, unexpected expenses, plus market drop, & now needed new job. Then serious thyroid and frequent need to go to the bathroom difficulties. MANY applications and resumes put out there, yet no interest. BUT, yesterday finally got a PT entry level food service job. It's a fresh start. Trying to look on the bright side.
Member since 8/6/2002

Aging increasingly attracts the attention of Web sites, not just on "sites for old people," as the site for the American Association of Retired Persons (AARP) has been characterized, but in arenas that serve broader communities. Interest in aging forums has rapidly increased as older Web users realized that the sliding, technology-driven economy might be comparatively harder on them and that they might find strength, either practical or therapeutic, in electronic communities. The large numbers of aging Baby Boomers who have been out of work in the soft economy commencing with the so-called dot.com bust support their initiative to gather around virtual hearths. For example, the number of unemployed workers in the United States aged 55 and over had jumped 23 percent between June 2000 and June 2001, contrasted with a 10 percent jump in workers ages 20 to 24.[1] Although hard numbers are sparse and perpetually outdated, trade publications and the popular press frequently have reported that technology industry layoffs have widely affected older workers, Baby Boomers especially.[2] In her article positioning the white-collar workplace of 2001 as the new "sweatshop," Salon.com's Katharine Mieszkowski quotes author Jill Andresky Fraser:

> In certain industries, like technology and Wall Street, by the time you cross that threshold of being 40 years old you're vulnerable. In fact, [there's] an expression in the high-tech community. If you're in your 20s, you're desirable; if you're in your 30s, you're expendable; if you're in your 40s you're unhirable.[3]

In January 1998, the U.S. Department of Commerce conducted the Information Technology Work Force Convocation, followed by a series of "town meetings" around the country in which various stakeholders voiced regional workforce issues. These town meetings included a refrain among mid-career and older workers and their advocates who cited barriers in the information technology (IT) labor market. The Commerce Department's report concluded:

> The IT industry is populated by many younger workers. Approximately 75 percent of computer systems analysts and scientists, and nearly 80 percent of computer programmers, are under the age of 45. Many managers in the IT industry are in their 20s and 30s and may be uncomfortable hiring or managing older and more experienced workers. A Network World survey of 200 readers with some hiring responsibility showed that younger network managers are less likely to hire older workers than younger workers. Almost half of respondents 20 to 30 years of age had never hired a person over the age of 40.[4]

The "Information Superhighway" is dotted with aging advocates who have formulated positions and who monitor the progress (or lack thereof) in the age/technology realm. Much of this content focuses on retirement and the construction of the so-called senior surfer. Web sites that publicize the plight and progress of these silver surfers include Seniornet.com, AARP.org, and Third Age.com as well as any number of federal and state Web resources, including the U.S. Administration on Aging and the National Institute on Aging. Less prevalent on these sites but growing in prominence in recent years has been content focusing on aging workers and technology. AARP.org, the Web site of the American Association of Retired Persons, has taken the lead here, but there are dozens more venues for content than even I had expected to find when I set out to do research in this area several years ago, including a considerable assemblage of Web sites originating from the United Kingdom, the site of new legislation protecting elders from a mandatory age 60 retirement. Because marketers and employers have finally discovered the aging of the Baby Boomers, scarcely a work-oriented Web site can afford to pass up a page devoted to the senior surfer or aging technology worker. I have found the richest such content in the user forums on these sites.

Web Sites Geared Toward Aging

Since I have been talking with people aged 50 and older about their Internet use, I have learned that their approaches fall into several groupings. While I expect that older users do distinguish themselves in some ways, I do not expect that they are hugely separate from younger adults in many of their Internet-related habits:

- For some, the networked nature of the Internet—its dissimilarity to a linear text most chiefly—deters interest and sustained attention. They either never get their "sea legs" or stay away entirely for fear of not being able to do so. This seems especially true of some of the oldest users, who generally are not involved in computer-related paid work.
- For others, surfing resembles joy riding: It yields interesting information but lacks direction. It is a bit like paging through magazines in a waiting room and does not result in sustained personal involvement.
- Some elders, after a period of exploration, settle into a ritualistic routine, having identified Web sites and communities of special affinity. This tends to be true both in personal settings and in work-related contexts. For example, when using the Internet to procure information or services, the elder might habitually go to one or two particular sites that he or she has found reliable. But this does not mean that the elder will not "switch brands," so to speak, if presented with acceptable reasoning for doing so. For example, I interviewed one man in his late 50s who had been an enthusiastic Travelocity.com user for making his travel arrangements at the office and at home. He began using Priceline.com for personal travel when he learned that he could save a large amount of money, but he continued using Travelocity at work because he felt that its more structured reservation system was significant there. Another man in his late 50s had selected www.aarp.org as his home page. This portal increasingly filtered his knowledge of the world as he supplanted television-viewing hours with hours spent on the Internet.

Job Seeking: The Major Net Theme for Over-50s

In 1999, the AARP Web site published "Working Options: A Guide for Midlife and Older Workers."[5] The document focused on the growth of the labor force and the (then) falling unemployment rate, based on 1997 statistics from the federal government. For its over-50 membership, AARP offers a list of internal links for information ranging from looking for work online to a discussion group called "The World of Work." Topics covered in "Working Options" include a self-assessment quiz designed to help workers determine if it is time to change jobs, "Refocusing Your Skills," "Use Your Homemaking and Volunteer Skills," "Handle Difficult Interview Questions," and other matters directly relating to tailoring later-life work efforts toward a satisfying direction. The Web page points toward optimism from all directions. Even when dealing with the dicey concept of age discrimination, the text attempts to empower its user, as in this example from the "Frequently Asked Questions" page:

> Q: *I am repeatedly passed up for promotions. Could this be age discrimination?*
>
> A: It could be a case of age discrimination. However, we encourage you to talk to your employer first. Perhaps you need to reevaluate your skills. Be sure you have the skills to match the job. If you still feel it is, a charge of discrimination must be filed with the Equal Employment Opportunity Commission (EEOC).

Although it provides clear directions for pursuing legal remedies, the site tacitly recognizes that age discrimination, legal or not, is to be expected. For example, on a page called "Write a Winning Résumé," AARP advises women who are going back to work after raising children to construct the document so as to disguise that career interruption. The site advises all mid-career and older workers to de-emphasize dates by omitting college degree dates and mention of earliest jobs and inserting time frames rather than dates of employment for relevant work experience that is in the distant past ("five years" instead of "1965–1970").

Likewise, the site warns job seekers to expect blunt questions from employers when they interview in person ("Handle Difficult Interview Questions"). For example, as users of the site learn, it is not unusual for unskilled managers to ask older workers how old they are. AARP gives this advice:

> Although this is not an illegal question, it is a poor question for an interviewer to ask. If you are 40 years and older, you are protected by the Age Discrimination in Employment Act (ADEA). Knowledgeable managers would not ask this question. But many managers are not skilled interviewers and do not realize that this is a question that could lead to a problem situation. If the interviewer asks this question and does not hire you, he/she needs to prove that you were not selected because you lacked the qualifications and not because of your age.

If you really want this position and feel that the interviewer has no discriminatory intentions, do not react negatively to this question. You might ask in a positive tone, "How does my age affect my ability to do the job?" Stress your skills and abilities to get the job done.

Job seekers get warnings about comments that might subtly mask age discrimination, such as a statements about apparent "overqualification" for the job or questions about whether they would be comfortable working for younger supervisors. AARP pragmatically advises workers to rechannel such questions into positive directions, to help the employer focus on the worker's skills, adaptability, and willingness to learn.

On its page "Overcome the Barriers to Employment," the AARP work site lists a set of myths associated with older workers and empowers the site's users to defeat these myths, as in this example:

Older workers are rigid, not as adaptable, slow to learn.

Consideration: AARP's research has shown that this perception of older workers is common among corporate managers. These traits [adaptability, ability to learn fast] are also cited by managers as those they consider most valuable in today's changing workplace.

Strategy: Dispel this stereotype by communicating your excellent record and expressing a desire to learn. Highlight new skills you have acquired (i.e., technology skills) and give specific examples of your flexibility and ability to adapt to change in the work environment. Point out instances where you have taken the initiative to suggest change and a creative solution to a problem.

On a different page, older workers are directed to "Pursue Lifelong Learning," with an emphasis on the need to catch up to changes that technology has brought to their work. "Retool! Retrain! Recareer!" they are advised, and information about distance learning opportunities and access to career learning centers in their communities direct them how. Presumably, most users who access this site already have some computer skills and probably an email account. The site goes into little detail about what kinds of training might be useful, and it might well function as a feel-good piece for users who have already become motivated enough to acquire a fair amount of skill in using new technology. The page has no emphasis on special challenges elders often face with new technologies, such as difficulties of keyboarding with arthritis-stricken fingers. It is reasonable to imagine this section having been produced with healthy, middle-class workers in mind.

What comes through in the AARP work site is a stubborn insistence that wise older workers will triumph if they can finesse perceptions of their worth in a world that must be understood as already discriminatory. Other axes of discrimination—gender and race, for example—are not taken up explicitly by the Web site, although older women are coached about submerging their identities as homemakers. Multicultural representations in photographs of individuals meant to stand in as mid-life and older workers dot the pages of this site, lending the expectation that everyone can overcome these equal obstacles in the over-50 job search. The site warns elders, when asked about their health, to acknowledge any visible disabilities and proactively note if they can do the job in question with reasonable accommodations. But there is little concrete advice beyond this.

Hopefulness, practicality, and, as a last resort, accessing the legal system combine to leave workers with a feeling of empowerment. AARP never warns them away from jobs as "too difficult" for elders. The logic conveyed matches that mythic American notion that anyone who works hard enough can become president of the United States. It is perhaps disappointing that, although AARP implicitly acknowledges discriminatory practices as routine in the workplace, it lays responsibility for dealing with these at the workers' door, advising them to contact the Employment Equal Opportunities Commission (EEOC). The site has no list of companies that have been found to discriminate. It contains no analysis of EEOC's success against offenders or any statistical prediction of the likelihood of older job seekers' landing jobs in particular lines of work. In essence, there is no willingness to take on the corporate status quo. Furthermore, in fact, no distinction is made regarding discrimination in hiring between for-profit and other enterprises. No sociological profiles are offered: Do white males over 50 seeking white collar jobs have a harder time landing jobs than Hispanic women over 65 seeking blue-collar jobs, for instance? Readers get no hint of how the finer cultural conditions in which they find themselves embedded will further affect their fortunes. The "world of work," under the AARP umbrella, is a homogenous one, presumably middle class, and "mid-life and older workers" are fruitlessly combined, the result being a bland list of generally helpful tips.

The AARP site appears to presume a middle-class surfer and takes no great pains to recognize important markers of cultural difference in advanced age. It normalizes late life by demonstrating that older workers, despite the challenges of age, are regular people, and they can survive in the contemporary world of work. An empowering message, to be sure, but one that falls short of recognizing that elders face diverse obstacles with varying chances of success. It is a variation on the myth of the American dream, and we are left to imagine that, if we try this advice and it fails us, we do not deserve to have succeeded. The outcome of such a logic is potentially damaging for older workers: While it teaches problem-solving skills and encourages agency, it does little to help people understand their special circumstances, and it does not take up the political importance of changing these conditions.

AARP's ample display of faith in the equity of the job market seems naïve when emanating from an organization whose job it is to question social policy regarding older adults, but it may be intended to be empowering for its audience. Such empowerment strategies, without concern for changing systems, risk setting up their users for failure. The phenomenon is a familiar one in media culture: the imperative to transform social issues into the realm of the personal, implicating the autonomous individual alone in the exercise of agency.[6]

Career advice elsewhere on the Internet largely gives off a similarly hopeful glow. For example, on the search engine site Altavista, the site's career site in late 2000 contained an article called "Ten Top Ways to Beat Ageism and Get a Job if Even Slightly Over the Hill" (*careers.altavista.com/ageism.html*). In acknowledging Baby Boomers as its audience, the article plays to the group's continued fascination with its own youth:

> Consider these ideas for staying in the employment game:
>
> Shave years off your looks. Get an evaluation from a salon and dress shop for makeup, hair and clothing. Or if you are a male, get an evaluation from a barber and men's clothing store. You may be putting out signals that are not with it. You may shave years off yourself by looking as up to date as possible.
>
> Stay physically fit: there is a distinction between biological age and chronological age. If you line up people who were born in the same year, you will see people who have aged gracefully and those who have not. Chances are those who look older are people who have not exercised continuously....
>
> Seek professions where a bit of gray is revered rather than reviled.
>
> Get to know younger blood. Sometimes people become outmoded because they choose to spend their time solely with people from their own age cohort, which can be very limiting. The wider the range of people with whom you spend time, the more receptive you will be to new ideas.
>
> Embrace computer skills. Let's face it: computers are here to stay. The more you know your way around them, the better.

Another trove of job-seeking advice can be found on the frequently visited Web site *Careerpath.com*. A four-part series entitled "Career Strategies for Older Workers" subdivides job seekers into those in their 40s, 50s, 60s, and 70s and beyond. Advice in each article is geared toward expected professional stages, family demands, physical issues, and workplace concerns for each age group, and this segmentation makes for a more realistic encounter than the AARP site allows. Job-seeking issues in one's 40s generally depart sharply from those in one's 70s. The advice is, unavoidably, still very general, but it does help the user identify special issues of significance that frequently are found along the fault lines of age, as in the third article, devoted to users in their 60s:

> Do co-workers and supervisors assume you will be retiring soon? If so, you may have to fight for the training and equipment you need to do your job well. There may be an assumption that you are no longer interested in adding skills, or that you're not worth the investment.[7]

Even in this fairly specific piece, the obligatory "take a computer class" hint gets imparted. As with most articles of its kind, this one contains no specifics: What kind of computer class? Toward what end? When older workers are told of the importance of updating their information technology skills, they must go somewhere else to learn how to get beyond this vague directive.

Careerpath.com also features occasional "Career Makeovers" on its site, such as a profile on 54-year-old Richard Davis, who felt his advanced age negated his career dream of moving from business consulting to a role as university professor. In the article, career consultant Melissa Martin insists that Davis's fear of age discrimination is unsubstantiated by fact that employers are delighted to hire older workers.[8]

One of the most upbeat pieces on careerpath.com's site celebrates the wisdom of age in an article designed to show how older workers can more successfully work with their juniors. In "Cementing an Alliance for the Ages: An Older Generation's Knowledge Can Power Your Career," aging Baby Boomers are instructed to quit worrying about their own obsolescence. "Although many of those twentysomethings are coming into the workplace with tech skills, they're missing one big thing: an understanding of how the workplace actually operates," the article concludes. Intergenerational communication would improve in the workplace if older and younger professionals would simply ask either other for help, the article concludes.[9] Such articles as this one assume a middle-class, professional audience of users and fail to unpack the class-based realities that sometimes muddy intergenerational communication in the workplace, such as questions of who is supervising whom, and how do these differences align with those of gender and ethnicity?

It may be overly romantic to imagine two office peers, one Graybeard and one barely shaving, shrugging their shoulders and sheepishly trading tips on corporate memory for computer memory. Much more plentiful are the kinds of working contexts in which people of color report to whites, women report to men, and, of course, unskilled and working-class workers of both sexes and various races and ethnicities report to salaried white males. In some cases, the salaried employees are older, senior workers in the company, sometimes occupying positions that are vulnerable to hungry attacks from younger (lean, mean) employees. In other cases, supervisors and managers are junior in age but well educated compared to the more experienced workers whose jobs they control. Asking for help under such conditions as these may seem fairly unrealistic to most workers.

As with much Web content that is administered by organizations, the text of these "tutorials" has a top-down tone. The interactive content of the message boards, by contrast, reveals a messy array of realities confronted by a diverse range of aging workers. In the balance of this chapter, I relate my analysis of four popular Web forums for older workers.

Aging/Work Forums

Between January and August 2000, I studied Age Issues on Monster.com, MSN.com's Aging Horizons, and the various message boards on aging and work issues both active and archived on Third Age.com and AARP.com's Money and Work message board. Much has been written about identity and presentation practices of people participating in listservs and other forums.[10]

I have examined approximately 4,000 posts to age/work-related message boards. Although I cannot take the content of the messages at face value and have little knowledge of its accuracy, I have made some choices about how to handle it as "data":

First, I take a phenomenological position: The content I analyze is part of a dialogue, a text, in which parties represent their *"lived experience"* in the world of work and age. Just as I will make no conclusions about the reception of the dialogue found on these lists, I will make no conclusions about the essential experiences the dialogue represents, only about the representations themselves. I am interested in the *construction of meaning and the discursive elements* of that lived experience, as they might be encountered to visitors to the message boards, rather than in validation of the events themselves.

Second, I am taking these postings as individual and community performances, in the vein of Erving Goffman's observation that all communication is a *performance*.[11] Participants on these boards variously construct themselves as experts, victims, agents, political actors, and many other roles. Not only do specific posters transparently perform specific roles to suit the occasion, but it is certainly possible that posters switch screen identities within particular lists or as they migrate to other boards. Performance of identity is context specific, as noted by other observers of Internet culture.[12]

I observed three common "rules" in the grammar of listservs and message boards devoted to aging:

1. *On these message boards, unlike other situations in which the participants often found themselves, age was at last not something they needed to hide from others, and in most cases, it was a point of pride.* As a refreshing break from much public dialogue, most people tended to volunteer their ages, and I had

no reason to doubt their veracity. I did notice that on some message boards, such as Monster.com's Age Issues, people posted their ages in a straightforward manner, rather than coyly, as some folks did on the *Third Age.com* discussion groups (e.g., "50-plus"). In all the message boards, members were able to maintain anonymity through screen names, although many dispensed with this protection, "signing" and addressing their posts, providing email messages, and reporting links to their personal Web sites. Some members clearly had gotten past "age as stigma," although others had not. In their posts, they reported, almost braggingly, that they "looked young."

A few posters went to lengths to draw attention to their embodied identity, as in the case of 60-year-old Bunzo's posting of her photographs (at home with her dogs) on the MSN Aging Horizons group site. While such postings may in part be a need to perform, to be seen, in the case of these users it seemed part of an effort to build personal closeness in their community. Only a few people, all women, posted their pictures; of these, the posters were among the most community oriented in their messages. Significantly, however, they were also "young looking" in comparison to their reported ages, and there may have been an element of status building among their peer group. Bunzo, having chosen a screen name that calls to mind a woman's hips, made a few onscreen remarks indirectly about her body. For example, when noting the revealing apparel worn in her office by a young woman, Bunzo mentioned that it was "one of those nothing tops that could only have been worn by those young girls who are small on top." Such cues seem to signify Bunzo's insistence that she has maintained her own femininity after its commodity value has been thrown into jeopardy, when her secretarial job was eliminated even though younger women workers were retained. Bunzo and others' self-presentations on the Web resonates with Marjorie Kibby's analysis of women's home pages, suggesting that women position themselves on home pages as significantly feminine and sexual, as contrasted to their more neutral practices of self-presentation in discussion groups.[13] The "photo albums" and "scrapbooks" that Bunzo and others like her posted as companion pieces to the message boards created a contiguous home page space for gender-role repositioning. Jamie Poster studied a group of lesbians who left their own chat room to visit a fundamentalist Christian chat room, where they held a "kiss-in." Poster skillfully described the tactics that chat room participants use to strategically exploit their absent bodies and bring them into the picture when doing so serves their purpose. As Poster concludes, virtual spaces present both problems and opportunities for people who are basing their identities on physical bodies, as in the case of LesChat.[14]

Shared status as troubled aging workers brings these communities together on the forums, but the community memberships are hardly cemented. Much of the dialogue on the boards is about how to escape their common situation, and members who succeed in losing the very thing that gives them their credibility—they get a job—disappear immediately, with few exceptions. Ironically, their loss of a certain kind of capital is a gain of a much more desirable one.

2. *Age discrimination, by far the most common theme found on the boards, was almost unanimously claimed to be the root of most older worker's problems.* Age discrimination certainly exists, and many of the posters provided compelling anecdotal evidence of its mark on their careers, but I had to wonder—particularly after "listening" to the posters for several months—if it explained all victims' stories equally well. Just as impressive to me was that fellow posters rarely pressed "victims" to perform any introspection. A tacit contract exists between participants on these boards: One accepts fellow members' explanations of discrimination, although individuals' reactions to that discrimination vary and draw diverse comment. Members go to lengths to observe their agreed-upon belief that discrimination against them on the basis of age is their great equalizer; as posts heap detail on this basic contract, however, ruptures occur, limiting a community's power to envelop members. For instance, tensions sprang up around differences over whether one can emerge successfully from a discriminatory situation or whether one should turn down a low-paying job because it is insulting to one's credentials. In computer mediated communities (CMCs), cohesiveness is tenuous. Always in question is how much of a community a virtual community, in fact, is.[15]

Internet-based communication, especially that involving virtual communities, is extremely context bound.[16] Members of the community have little to go on when formulating opinions about fellow members. They may tend to "buy into" another's explanation in a more wholesale way than they might if the community were physically based and more cues were available as "data." Generally, until a user revealed something to the group to cause others to doubt his or her version of reality, the group allowed the user to place blame for all problems at the door of age discrimination. When group members seem to lose faith in the "blaming" individual's truthfulness or self-awareness, they are quick to dismiss much of what the person claims in future posts and chide the person for his or her failure to take responsibility.

3. *In taking one another seriously, participants demonstrate how elders might be taken seriously in a more equitable society (even if they have their squabbles).* Age is more or less invisible on the Internet. Despite occasional self-references to "gray hair" or "wrinkles," apart from the frequent acknowledgement of years of age, it is difficult to distinguish the writers of the posts on these boards as a cultural group of "elders," either so-called Matures or Baby Boomers. These writers' words do not play to type. They do not move slowly across the screen but appear as unified narratives, written "quick and

dirty" for the moment of consumption. They do not convey the notions about passivity, technophobia, decay, homogeneity, and foolishness that we so often have seen associated with elders. Nor are they splashing in the pond of some ridiculous second childhood.

Mike Featherstone has observed that computer network interactions "would seem to be ideally suited to the needs of the elderly faced with restricted mobility, impaired communicative competence and other "bodily betrayals."[17] There may be a vast world of difference between the speed and mobility of the youngest Baby Boomer and the eldest "Mature" Internet user, but the technical disembodiment of the users on these message boards encourages a shared generational status to the extent that they self-identify as "aging workers."

The Aging/Work Message Boards

By far the most solid "community" among the four message boards I studied was the Monster site, and I will talk about its structure in detail here. I will follow up with briefer descriptions of the other three sites to show how their communities distinguish themselves.

Monster

Monster.com, easily the most popular of the World Wide Web's growing number of jobs-related sites, hosts twenty-eight (accurate as of January 2003) member forums, including one called Age Issues. Like many interest forums, the Monster Age Issues forum features a blend of "regulars" whose posts conformed to theme parameters but functioned socially as well. Board regulars include people in their 60s and 70s as well as people in their 20s and 30s, but the forum is overwhelmingly a Baby Boomer community, with 40–55 as the predominant age range. Lest message board readers begin to forget this, posters liberally sprinkle their messages with quotes from such Boomer-era lyricists as Carly Simon, Bob Dylan, and, yes, Lou Grant.

The Age Issues message board functions as a community for some and as a temporary resource for others. As with many forums, a large number of authors posted only one or a few messages and either found the forum not to their liking or moved on because their needs were satisfied elsewhere, sometimes through success at finding employment. When "regulars" are in the midst of a thread in which they are discussing one or another's personal issues, a high volume of venting, reassurance, and calls to activism takes place, punctuated by occasional "flaming" from unsympathetic members who report being tired of the venters' "woe is me" attitude.

When such exchanges have been silent for a few days, Monster visitors sometimes wander onto the message board, generally issuing brief posts describing a recent job loss and possible search that is going nowhere and asking the Career Mentor or others for practical advice. Response usually is liberal and discussion tends to remain on track, with some "regulars" tending to offer experiential advice by capsulizing their own stories. Occasionally the "newbies" stay, but more often their postings stop after Age Issues resumes business as usual, with regulars reporting on their situations and fellow members reaching out to them. Sometimes a board member offers a sympathetic reassurance but sometimes instead deals out a "Get a life!" sort of response, reporting months of "lurking" without having previously posted. Something in the "regular's" post moved the "lurker" to respond, and, after a quick exchange between the "regular" and "lurker," the latter is not heard from again.

With the frequency of about once every two weeks or so, a "regular's" post so irritates another, confrontational regular that flaming ensues. Invariably, a handful of the same regulars send a quick barrage of posts to reset the group's equilibrium with some condemnation of the flaming practice by "mean spirited" posters who have no right to "pass judgment" on another's feelings and with sentiments for the person who has been flamed, reassuring her/him (usually her) of self-value and encouraging her/him to "hang in there."

Of the twenty or so regulars, most of the vocal ones are women, and among these a few frequently report on their personal sagas in great detail. A core of about half the regulars, mostly (but not all) women, remain devoted to these discussion threads, and the balance of the group, mostly (but not all) men, participate in distinctly separate threads. More often than not, when a thread is authored by people identifying as one gender only, it is male.

Race is hardly ever mentioned; when it is, it is usually in the context of the compounding of "-isms" that a poster has encountered in the workplace or job market, including age and gender biases. Gender bias complaints air frequently. Some liberal-minded male posters join threads surrounding this issue, but most males steer clear of them.

Because the forum clusters around a single issue—age—differences over race and gender perspectives, as well as over generational (Baby Boom versus Matures) and class (working versus professional) perspectives are apparent in the message board but go largely unengaged. Posters generally find threads of interest to them and in which their views are largely validated, and they tend not to participate in ones that do neither. Mia Consalvo has theorized that the increasing segmentation of Internet users, in the mode of market segmentation, constitutes the formation of "media streams," which are, as they sound, fluid constructs that anticipate and respond to dynamic technologies and social environments.[18]

In this secular space, religion is almost never mentioned. Neither are preferences within the two-party political system, but some do criticize its narrow practices. For instance, in 2002, one participant frequently hails Ralph Nader as a political hero, and there was one endorsement of Pat Buchanan. Both politicians were touted for their promise to disrupt business as usual in Washington. Several participants frequently criticize national and state political systems for their lack of empathy for the older worker, but posters take diverse positions about the issue, with some condemning the practice of "sending our jobs overseas," relegating the older worker to obsolescence, and others espousing a more radical view of status-quo politics for supporting discrimination against the aged, women, and people of color. Some regulars reveal in their posts a well-developed understanding of capitalism's costs and benefits, while others sound almost paranoid about people they see as conspiratorial legislators and corporate figures who are "out to get everybody over 50." In this very heterogeneous political forum, it is not surprising that threads on these matters stop short of much development.

The status of "Career Mentor" Linda Wiener never emerged clearly throughout the six months during which I observed the listserv. Wiener posted approximately weekly and did not function as list moderator. Sometimes participants would query her by name, and she usually responded, but sometimes she was apparently absent. Once, when she did weigh in on a query, three board members quietly flamed her for not responding more frequently and for not providing more direct information in place of links that were relevant to topics that had been raised. Although she did not respond directly to this criticism, Wiener wrote more detailed personal responses in subsequent posts.

Wiener, whose Monster biography reported career mentoring for aging adults through such organizations as AARP, responded only to posts that were fairly generic and provided her with a stage for a "teachable moment" for others who might be "listening." Although she occasionally welcomed newcomers to the list, she never publicly acknowledged the personal business described in her core members' posts, as when one woman acknowledged that she was contemplating suicide due to her desperate economic crisis. Others in the core group were quick to fill this gap and provided sound advice and sympathetic comments. Their efforts at solidarity tend to drown out the cynical sniper whose answer to those who report that they are in dire trouble is effectively advising them to jump. The critics apparently are either bored with chronic whining or suspect that the whiner is merely seeking attention. The harsher critics tend to participate more constructively in threads that emphasize practical action in job hunting.

Monster.com Age Issues Forum Participants   Monster's community enjoyed the largest degree of individuality. Seven regulars demonstrate the extent to which this was shown.

Jenn2, 49, reported having some type of IT degree and more than 10 years' experience in the field of information technology. Having been laid off three times in four years in the Boston area and subsequently out of work for almost two years, she collected unemployment insurance benefits for most of that period, apparently, and served for months as the message board's unofficial, self-appointed unemployment news correspondent, providing link after link to journalism and government Web sites as Congress deliberated on extending the benefits period during the current recession. At one point, she reported turning down a job offer in her field from Microsoft in Massachusetts, a move that sparked an onslaught of posts from chiders who seemed to take personally her decision to remain jobless and penniless, like themselves, when she could be back in the world they still saw as "normal." Jenn2's defenders, unprompted, came to the fore, blasting the critics for their insensitivity and presuming that Jenn2 had turned down Microsoft because she was shell-shocked from having been laid off multiple times in this unstable industry. By June 2002, she had turned down a $27,000-a-year store manager job, then accepted a $35,000-a-year offer from Cumberland Farms, the convenience store chain, quitting four weeks into her training because, she said, she had been forced to take a lower ($8 an hour) pay rate for the training period and the work was exhausting. Jenn2's perspective on being laid off, having to fight for unemployment benefits, and having to take a bad job instead of a plum one for which she felt well qualified was colored by her certainty that she had been continually discriminated against as an aging worker. She served as a near poster child for some regulars but angered others with her "holier than thou" refusal to engage in what they saw as honest but low-paying work at menial jobs that would allow any of them to make a living or come closer to doing so.

Jenn2 was the one regular who occasionally sparked discussion about herself in the third person. Her harshest critics blamed her for assuming a permanent victim role and not exercising a rational outlook. I may be badly off, they seemed to say, but at least I'm not handling it like Jenn2. At times, after the sharpest attacks on her decision making, Jenn2 stayed away from participating in the message board for several days. She also avoided participation when newcomers introduced practical questions to the group. Days later, she would file a breezy post about unemployment insurance, a link to a news story coupled with a sarcastic condemnation of big government or big business. Usually, a regular thanked her for the enlightening article.

GettinSerious: Living in southeastern Wisconsin, GettinSerious was a divorced mother of a teenager. She had accumulated ten years' experience in information technology before going back to school and obtaining a doctoral degree in philosophy, which she taught for ten years prior to being turned down for tenure at two institutions and returning to the IT field. Downsized, she resorted to flipping burgers at McDonald's. Her biggest dilemma during the time I audited the board was whether to trade her $7-an-hour shift job for an $8-an-hour Arby's shift manager position, easing her financial situation but perhaps sealing her permanent departure from IT at age 49. She elected to go with management, but not before taking exception to Jenn2's inference that she was settling for a job that was beneath her.

Gettin' had difficulties sorting out her own situation and occasionally vented over her own frustrations. For example, she expressed extreme anxiety over having opted out of unemployment benefits for the privilege of supervising the service of fast food, hard and demeaning work for approximately the same cash. She felt foolish for someone of her years and educational attainment and bravely shared these sentiments with her peers. This acknowledgement starkly contrasted with Jenn2's lack of humility.

Gettin' was also among the readiest supporter of peers who expressed a need and a frequent advisor of newbies seeking practical help.

Moodusbass, 44, lived in Connecticut and had been downsized from an IT position two years prior to my joining the message board. With two large dogs, a 22-year-old girlfriend who had a 5-year-old son, and no replacement job, Moodusbass was going through bankruptcy and foreclosure proceedings, and frequently posted to the list with desperate ramblings and queries for advice. At one point, an uncle and aunt in Ohio had offered to allow Moodusbass to move into the home of a recently deceased grandmother. The group sorted out the issues surrounding this potential move, including trading privacy for the privilege of living rent free in a hometown that lacked economic opportunity. At one point in this discussion, Moodusbass corrected a poster's presumption that she was a male and, with some coaxing, came out to her fellow board members, to several members' approval.

It was Moodusbass's veiled threats of suicide that evoked sympathy, practical advice, and even the harsh "go ahead" comment from fellow participants. Her posts offered no opportunity for mutuality. They seemed to drain energy from the group. Toward the end of my observation period, Moodusbass reported in a "good news, bad news" post that she had landed a full-time job as an "engineering aide," in which she could use her computer training and experience, but that her recently departed girlfriend had started seeing someone new after only three weeks.

Bunzo, 60, seemed to consider herself the wise owl of the message board, having "seen it all." She rarely focused on her personal situation at length but appeared to enjoy helping others through commiseration and practical advice. Downsized herself in her late 50s from an office administrator position, she played a regular role on the board in maintaining a sense of continuity and camaraderie. In the midst of my observation, Bunzo let slip a few times that she would be turning 60 in June. Finally she mentioned in an offhand way her birthday's date, and when the big day arrived two regulars posted well wishes on the board, to Bunzo's delight. Bunzo and others also made note of holidays over the months: Easter, Memorial Day, and the Fourth of July. Holiday messages from Bunzo and others praised the greatness of the American ideal as they perceived it. In one such post, quotations from Thomas Paine and other "great men" reminded the community of its patriotism in spite of the hardships that had brought them together. Bunzo, older than most of the regulars, unabashedly posted "corny" messages compared to those of the ironic and sarcastic Boomers.

Bunzo's sincerity was always well received on the board, even if the Boomer members might have considered themselves comparatively hip. She embraced fellow sufferers unconditionally and, although her messages were generally sweet in tone, she quickly jumped in to rebuke posters whose messages she felt were mean spirited to others.

It is difficult to read the posts from these four women and not feel some sense of divided reaction to their collective plight. They seemed to suffer from varying degrees of depression, with the extreme being Moodusbass, who reported going through the loss of job, home, and relationship more or less simultaneously. The indignity and desperation she must have felt associated with her bankruptcy proceedings, disclosure about being gay and also jobless to family and friends, and uncertainty about where she would soon be living are indeed circumstances that most of us have never faced, and it was easy to fear for her and to empathize. Yet I came away with some concerns that if these women largely relied on the message board as a venting outlet for their feelings of discrimination, then they would be unlikely to escape victimization.[19] While others on the board actively sought routes upward, these women's self-disclosing posts seemed to grow more poignant each time, as if they were spiraling into further decline through the message board. As a group, they were by far the most frequent posters, weighing in for holiday sentiments and personal milestones with great dependability and at times with the tone of a diary entry. These women simultaneously hooked and repelled me, and I continue to follow the message board largely because I want to know how things are turning out for them.

Three men were also frequent posters, and their situations represented diverse fortunes.

Wristtwister eventually reported (on his own message board, MSN's Aging Horizons) that he had been banished from Monster's Age Issues forum for having touted his own site too many times. Victim, again. Yet, by the end of my observation period, his posts on Monster's board continued. Wristtwister's messages were laced with sarcasm and anger over ageism and what he considered to be the wholesale giveaway of American jobs to foreign laborers. Wristtwister was unafraid of controversy and confrontation. At one point, a 36-year-old poster told board members that he had tried to get into the U.S. Army but was told he was too old to join. Another poster validated the army's position, noting that some jobs have physical requirements that aging workers cannot meet. Predictably, wristtwister jumped in:

> [A]nd we don't have any military jobs that aren't physically demanding that a fit 40 year old could handle? What about training? Want to pit one of your 20 year old wonders against my 40 years of martial arts training.... I'll win... guaranteed. Age discrimination is most prevalent in government service... the same government that is supposed to stop it, and that issues the guidelines so business can practice it with impunity... and by the way, the same government that WE pay for. Of course you can't sue them, because they've legislated themselves immunity, so until we figure out how to force the government in a republic to do right, we'll just have to live with what we've got. Personally, I think it sucks...

Crucially, for wristtwister and many others on this and other boards, "the government" and "America" were separate entities. He was pro-America even if he was anti-Washington. The government that wristtwister appeared to idealize was conservative, pro-white "American" labor.

Dirtman, a 57-year-old soil scientist in Arizona, was one of the few active participants who reported being happily employed. He never dominated discussion but frequently weighed in with dispassionate, rational advice for people whose questions were practical in nature. He considered himself a "success story" who had found happiness after unemployment and stayed around to help other aging workers through their dilemmas. Negative talk was anathema to Dirtman. He never acknowledged "victim" talk on the board and always encouraged others to turn their dreams into rational steps. As his user name indicated, he was "down to earth" and championed the strategy of scaling down one's economic expectations in lieu of staying in the "rat race" at too high a personal cost. Dirtman liked to point out that people can live on modest incomes and achieve happiness.

Marketing Commando, another success story, frequently contributed to the message board before suddenly disappearing. Several weeks later, Marketing Commando resurfaced on the Monster board, and continued to post upbeat, strategic advice to fellow users. The transient nature of message board posters is a defining element of Internet communities.[20] A poignancy frames such appearances and disappearances on the job boards. The arrival of a poster often carries with it the bad news that the participant has lost a job, is having trouble finding replacement work, and likely is feeling age discrimination's effects. Often, the aging/jobs board's fellow travelers express their collective sympathy with the reassurance that "we're all in the same boat." Small comfort, perhaps, but such reassurances tend to convey both comfort and validation for members of the existing community, whose suffering likely has far exceeded that of the new poster who has been out of work a mere two months or so. The community struggles to express such mixed feelings.

When a poster suddenly exits, as the apparently "successful" Marketing Commando did, without showing up in a contiguous community, such as another board on Monster.com or on another aging/jobs board, little if anything is expressed in the community to acknowledge the departure. I cannot help wondering what those left behind might be feeling. Does such a departure constitute further abandonment of these individuals who have lost out on so much already? When the Marketing Commandos of the boards leave without a word, have they thought to themselves, "Glad to be out of there!"

MSN's Aging Horizons
This site is managed by wristtwister, the Vietnam veteran/martial arts instructor who frequents the Monster board. Although many other men also post to the board, wristtwister's regulars mostly are women. On the whole, the Aging Horizons board skews a bit older than the Age Issues Forum, although it includes some of the same regulars—Bunzo and GettinSerious, for example. Notably, the sometimes marginalized Jenn2 and Moodusbass from Age Issues are absent. Content on the board is more politically strident. Wristtwister reports to users at one point that he has been "banished" from the Monster site for having made too many references to Aging Horizons. His response on these and other occasions—and indeed in posts that he and some others have made on the Monster Age Issues Forum itself—is that Monster is timid about taking on some of the more political issues in its forum, because it is a profit-making enterprise that cannot afford to offend the companies that advertise on it.

The Aging Horizons board has a special list devoted to "Immigration," where the majority of posts are from wristtwister himself, who often shares links to Web sites of interest (mainstream and special interest). He sometimes uses "editorial license" to cut and paste Web-based stories into a post. On this list, xenophobic rants condemn a mound of U.S. government policies and practices that the posters see as having resulted in stacking the employment deck against them. Several regular posters to this list and others on the board stop just short of hate speech as they express their "patriotic sentiments" in this venue. For example, Dee Bryson, Environmental, Health and Safety Manager, who also posts under the screen name Ragdolkid on the AARP forum, filed this post on Aging Horizons' Immigration list:

> TEXAS, NEW MEXICO, AND ARIZONA, MY STATE, ALL HAVE MORE THAN THEIR FAIR SHARE FROM OVER THE BORDER!!! AND THOSE IDIOTS WHO WAS COMING INTO ARIZONA, AND GOT SICK, SOME DIED, WELL THAT IS THEIR FAULT!!! NOW THEY ARE SUING OUR STATE BECAUSE WE DID NOT HAVE THE WATER AVAILABFOR THEM, OUT IN THE DESERT!!! WHO DO THEY THINK THEY ARE!!!!!!!! **GIVE ME A BREAK!!!!**

The Aging Horizons list has comparatively little of the sustained personal narrative making seen on the Monster forum and also rarely addresses practical issues of job hunting. Overwhelmingly, the site is politically charged, filled with bitterness and a narrow sense of patriotism, and focused on references to the downturn in the IT industry and on Web sites that underscore claims about ageist employer practices and the like. In addition to a "General" list and "Immigration," other lists focus on "Discrimination," "Veterans," and "Disabilities," the latter having been established in response to a poster on Age Issues Forum who had been unsuccessful in getting Monster to set up a board devoted to disabilities and work. Wristtwister set up an "Advertising/Jobs" list as a way of handling the inevitable traffic from posters who want to hire or sell the promise of a job to the unemployed community members who monitor boards such as his. Many such posters to sites like this have a multilevel marketing (MLM) premise, which is the subject of both desperate interest and angry condemnation by regulars of the boards. Other board monitors handle the "advertising" posts differently: The Monster community moderator deletes them soon after posting and provides a form explanation that the deleted post violated message board rules. AARP's rules seem to be more liberal. Although some advertising posts are deleted, many are worded blandly enough to escape the monitor's delete key.

This exercise in penalty avoidance provides further fleshing out of the new rules for inhabiting identities in cyberspace, the new sphere of the virtually disembodied communicator. Alluquere Roseanne Stone's work on cyborg bodies and boundary encounters helps me develop the idea that business people whose everyday identities might be questionable or lacking in resources, evidentiary in conventional encounters, are testing the Internet to inhabit online communities where they might eclipse the constraints that have held them back so far.[21] Peter Lovelock and John Ure have observed that "electronic commerce is being shaped by, and increasingly will help to shape, modern society as a whole."[22] To this end, such experimental encounters between business and workforce as seen on the AARP Web site through the mechanism of a new mode of "advertising" indicate that the Internet's economic cores are still quite liquid. New economic models and communication practices, as witnessed in the struggle for media corporations to make a living off the Internet through such practices as banner advertising, are still very much under development.

ThirdAge.com

ThirdAge.com is a cheery celebration of active aging. Users encounter a portal replete with advertisements for hair dye, denture cream, and heart medication; news for "activating your days," "finding your sexual peak," and "keeping your family on speaking terms"; interactive sections such as "Vote and vent" and "Recipe finder"; and, of course, discussion groups. These include groups concerned both with technology and with money (vis à vis retirement income) as well as a site devoted to Work, which includes several active and archived lists: "Creating Meaningful Work," "Job Hunting Over 50," Work Styles for a New Millennium," and related titles.

Community presences on these lists tend to be short lived and infrequent, with most posters attracting only 100 or fewer posts before retiring into the archives, which remain accessible online. The most active list at the time of my research was "Work Styles for a New Millennium," which contained some of the give-and-take downsizing narratives and strategy sharing that are commonly found on the Monster board. Most posters are Baby Boomers over 50.

The site's distinction is its New Age–like theme, with both some list moderators and many of the users talking up strategies for "empowerment," "synchronicity," or "the Zen of work." This institutionalized spirituality provides a close match to what Tara Kachgal has identified as "commodity spiritualism," which she found as a rhetorical device in Nike's Web site NIKEgoddess.com, part of the corporation's "Goddess" marketing campaign.[23] It seems that list participants mostly agree that they will make the most out of a bad economy, given the constraints of what they are sure is an ageist culture.

AARP

AARP's boards are perhaps the least like a community. This seems to be true not only because posters have the prerogative to sign in as "Guest" on the boards, and about half do so, but also because AARP is such an imposing presence. When Internet users visit Monster.com and wander onto one of its message boards, they are quite liable to perceive themselves as visitors, short-termers, who are there to problem solve and who expect to move on. On the other hand, when Americans turn 50, send in their AARP dues, and check out the Web site, they have joined up, quite possibly, for life. They are no doubt aware that AARP is a giant organization, a household word, a powerful lobby. The second level (beyond a screen name) of anonymity they derive from signing in as "Guest" hints at their need for some protection from the Big Brother they have connected to.

Some posters, even several who sign in as "Guest," do add a screen name, and dialogue between and among people does emerge, but it generally is less focused on personal connection. Individual tales are told in much the same manner as they are related on the Age Issues Forum, although the Jenn2s and Moodusbasses do not emerge. What does emerge is biting political comment, much of it a conservative critique of both the two-party political process and corporate bullying of "the little guy." One of the posters from Aging Horizons, Ragdolkid, frequently posts there, and much of her comment is anti-ageism by way of being anti-foreigner. On this list, Ragdolkid's posts are in all capital letters and in bold, as they are in the MSN forum, but they are also in bright colors and varying sizes of oversize, playful fonts. Her posts remind us that she is, somehow, Raggedy Ann. After one particularly lengthy post, a sarcastic critic replied: "I didn't quite catch what you said. Maybe you could use a larger font next time."

Other members of this message board are also playful in their identity construction, choosing names such as "Magnolia" and "Swampy Gator." These playful screen names are sometimes linked with sweet sentiment, as when "Swampy Gator" compliments a poster named "Kitty" for "a real good post, Miss Kitty." Age-related unemployment is the most popular topic, and it is often coupled with political rhetoric about immigration and free-trade policy and the perceived underprepared nature of young workers.

Aging/Jobs Community Themes

Among the four online communities I studied, I identified ten distinct themes. Almost all of the dialogue in these forums dealt with one or more of these. I have delineated the themes here in rough order of prevalence, noting that in some communities certain themes were much more prominent than others, and differently developed.

Themes tended to interconnect. For example, personal narratives often revealed a story of age discrimination in a slow economy and featured elements of venting. Posts about politics tended to center on discrimination. Strategies offered for job seeking touched on hiding one's age, becoming more technology literate, or referring to Web resources.

Posters tended to show partiality for particular themes. Some regulars posted often when a favorite theme was the topic, or they found a way to tie in their favorite theme in a topical way. For example, for wristtwister and Bunzo, the sharing of economic news predictably tied into concerns about a perceived political threat from government's favoritism toward foreign workers. Analysis of the themes follows.

1. *Age discrimination in downsizing/layoffs:* The unifying impulse to blame that joins almost every thread of each of the boards. Tacit or explicit, it is personalized through narratives of experience, cited through antidiscrimination media links, critiqued through media links identified by members as naïve, and discussed philosophically, strategically, and fatalistically. Overall, the mood is pessimistic, but strategic and pragmatic posters frequently seek advice and offer ideas on how "the little guy" or "gal" can triumph over this massive evil.

Personal loss narratives almost always carried with them some "proof" of the dominance of age discrimination in the workplace, either in hiring, among co-workers, or in firing or laying off employees. Some posters attempted to work through underlying reasons for the discrimination, such as cultural perceptions that older workers cost more in salary and health and pension benefits, do not learn or work as fast, and take more sick time. Almost universally, such perceptions received strong counterintuitive exclamations from the communities' members.

Posters frequently expressed these sentiments in connection with their views on age discrimination: It is difficult if not impossible to prove, because discriminating employers avoid admission of guilt, experience is devalued by contemporary human resource professionals, and bitterness is impractical. Such conclusions can be found in the post of Posey 18, who writes to the AARP Working Options message board:

> In 1997 at 59, in good health and two associates degrees in computers under my belt. I began to look for a job. I was turned down again and again. Until it dawned on my I was being discriminated against. But what good does it do to report it—it is seldom looked into by authoritiese that can do something about it.
>
> My case, I am married but my husband had to take disability retirement and I lost my health Ins. Through that. I took out health ins. As an individual and the premiums have called for us to have to cut other things to the bone to pay them.
>
> Our culture worships being young. And as a result older workers who are in good health and qualified are not looked upon as being the asset they are. Plus the health ins. Business doesn't like insuring older workers. Its a catch 22 if there ever was one. But there has got to be a way to make it work, maybe working at home?
>
> My dad was laid off at 50 something years ago—couldn't find a job no matter how hard he tried. And he went into a business for himself. His business went over great for him and it ended up to be a blessing in disguise. So remembering him I try to look for what fate may be trying to tell me. If anything.

Posey 18, like many others on the boards, draws together a string of popular assumptions that fit her experience. Claims such as the culture devaluing the aged and insurance companies not wanting to insure them are readily believable. Therefore, we can fairly easily buy Posey's view that she was discriminated against. We know little else about her (other than that her emails are filled with sentence fragments). What kind of job candidate does she make, apart from the cultural obstacle of her advanced age? It is not the function of the message boards to determine this.

Occasionally, a young lurker on a board would come forward and make an argument that workers of, say, under 25 also were targets of age discrimination by employers. Predictably, such statements attracted little comment other than brief dismissal from a veteran poster, and the young poster was not heard from again. It was heartening that the young interloper was not flamed. The posters tended to share a view of discrimination that stopped short of blaming others for being young. It is possible that some of the participants did not have to look far to see that the "youngsters" were close in age to their own children, some of whom were still being partially supported by the community members due to these young adults' difficulties in living on their own in a challenging economy.

The stories posters told conveyed a composite of people who had lost not only jobs but often a share of their self-esteem, crucial health benefits, and, if they had accumulated them to begin with, their nest eggs. Generally, such posts revealed feelings of betrayal, desperation, and need for serious advice on how to proceed.

Here is a typical post, from AARP's Working Options:

> About 30 years ago I entered the telecomm computer field and have worked my way up to Network Engineer at one of the top telecomm carriers—yes, without a degree. Through all these years I learned all the high profile systems.... I.e. switches, routers, ATM and frame relay networks. I trained many newcomers and was considered a key player with strong performance reviews.
>
> Well, about one month ago I was laid off from this company with two other individuals—cause work force reduction. Prior to these layoffs about six months ago an additional two employees were also laid off from this employer. We, the five employees, all belonged to a networking group of approximately 16 engineers and what did we all have in common? We were all above the age of 44 years old with several of them with the company for more than 5 years.

The poster sought advice on hiring an attorney and whether others thought he and his co-workers had a strong case against the telecommunications firm for age discrimination. In such circumstances, fellow posters tended to report their personal experience with lawsuits and complaints and to encourage the aggrieved poster to proceed.

Many such posts cropped up on the boards I studied. Often, posters such as the man whose story was just relayed reported that they had dedicated many years to a single industry and firm, only to have their economic world cave in several years prior to their being able to afford to retire. Men and women shared this general experience, but their narratives differed in an important way. The men tended to convey the sense that they had been stripped of some element of their identity—as bread winner, as professional, as *man*.[24] Some reported regret at having entered a field (some information technology sector) that disserved them so greatly during what they believed should have been the height of their working years. It was clear that all of these men were white-collar workers but that none was extremely high ranking. To them, the big corporation remained fairly faceless;

rather than lashing out at individuals, they did so at "the company." The women, who often were divorced or widowed, reported genuine fear about having almost no cushion against poverty after the loss of what already was a low-paying job. The women were much quicker to acknowledge that they would be looking for work at McDonald's or in some other low-paying capacity. Both men and women reported seeking and finding work at temporary employment agencies, although this was more often the case for women.

Through descriptions of their job searches, many of these men and women conveyed a grimness about the unlikelihood that they would be able to find comparable replacement positions. Although corporate downsizing has been a factor in the American economy before, such as in the early 1990s, the 2001–02 recession, compounded by the post–September 11 downturn, aggravated such loss totals, and members of these message boards acknowledged that complication often. The vehicle of their anger was, surprisingly for me, not the Taliban but big corporations, with their greed and ethical breaches, and the federal government, with its perceived apathy about their personal losses.

2. *Venting, flaming, and reassurance:* Community and competition permeate episodes of venting about age-related vicitimization on these message boards. "Okay, you're not going to believe this one, folks," or similar statements, tend to open such narratives, and the reader is invited to both identify with the poster's victimization and acknowledge the drastic degree to which it has been experienced.

Victim status is closely linked to age discrimination, but it is widely rejected by pragmatic board members whose concerted goal is to get on with life. A diversity exists in the community, and that diversity extends to levels of tolerance for self-pity. Generally, most posters seem to realize that, if they convey a sense of wallowing in self-pity, they will lose their sympathetic audience, or worse—they get flamed.[25] A few posters, most of them on Monster.com's Age Issues Forum, seem either not to realize that they have crossed the fine line between venting and wallowing, or they do not care. Moodusbass and Jenn2 are the most extreme examples, and both frequently air posts that attract a small number of reassurances and a few condemnations from flamers. Most regular posters tend to ignore them both.

GettinSerious, the Wisconsin Arby's assistant store manager, vented in both the Age Issues Forum and wristtwister's Aging Horizons community when she realized that she had been caught (or caught herself) in the trap of hard work for menial pay when she could have been collecting unemployment wages over the same period and spending the time looking for a more appealing job. No one answered either post.

Some posters who were worried about age discrimination were surprisingly young. Understandably, 18- to 25-year-olds complaining to the list that they had been discriminated against by employers who said they were "too young" for jobs did not arouse deep sympathy from board regulars. Neither did 30-something posters who were worried that they would be seen as too old by employers. One young man, Bt54, a 32-year-old chemical plant worker with a high school education, a stay-at-home wife, and two small children, did manage to sustain the interest of board regulars through a lengthy thread in which he posed questions about his own career crossroads. He wanted to get a college education in order to seek white-collar employment in the earth sciences, and for several months he queried the group about his options. He took the spectre of age discrimination seriously. His biggest dilemma was whether to work full time and take one or two courses at the local state university each semester or to accept his in-laws' offer to move his family in with them and go to college full time. Ultimately, he heeded the group's warnings about giving up his privacy and control and decided to earn the degree while working, but he worried about entering the job market at 40-plus years of age. Group members took seriously his concern that he might be subject to ageist treatment at what was for many of them still a young age, but they encouraged him to go ahead and shore up his future through pursuit of a degree. He left the board still expressing some uncertainty but seemed reasonably at peace with his decision.

3. *Political knowledge, opinion, and activism:* Heavy traffic accumulated around a desire to have unemployment benefits extended and anger over perceptions that older Americans' jobs were being deeded over to foreigners, either inside the country (immigrants) or outside it, via free trade agreements and corporate money-saving tactics. Wristtwister's Aging Horizons on MSN.com had an entire sublist devoted to immigration, and he dominated the dialogue with frequent posts, often citing alternative press and Internet content that attacked U.S. immigration policies. Posters on the general message board of Aging Horizons, most of whom are women, were quick to express sweet sentiment and sometimes affection for one another. On the "Immigration" list, the posture became harsher. Both Bill Clinton and George W. Bush received condemnation in such postings for their "opening" of the border to Mexico.

The desperation that some of these workers felt because of their career situations serves an apparent catalyst for bitter venting when they share news and opinion about foreign workers as interlopers seizing their benefits, as in this comment from gloryglory on Aging Horizons:

> Well, it's happened! The illegal immigrants stranded in the U.S. who've been sending their wages overseas for x years are jobless too. But get this—agencies are paying their rent and giving them food money. http://www.villagevoice.com/issues/0143/lee.php

The few times I asked public agencies, paid my your tax dollars and mine, I got the accusation of "laziness", that I could get a job anywhere. What happened was illegal east Asians, who were told by the disaster relief people that they could get a job anywhere, couldn't; but there was an agency just waiting to help them stay here until they could resume sending money overseas.

I've been tramping this road for almost 20 years. I'm convinced there is no help because of age—even those that think they look young just attract nasty comments about trying to be someone they aren't.

The only solution I have found is to enjoy life as it is, whether or not it's fair. There's always some cute, helpless person wandering around sucking up the handouts, and I was never cute.

Gloryglory, as with other self-identified victims on the boards, takes personal offense at the small successes of other kinds of victims rather than questioning the systemic (governmental and market) forces that leave both herself and foreign poor workers at a deep disadvantage. She and several of her co-posters on Aging Horizons, as well as on the AARP message boards, come close to constructing "hate group" politics, as if the rendering of others as pariah will somehow ease their own economic plight. This is similar to way that poor whites will express hatred toward blacks to feel better about their own social position.[26]

The Web site sponsors themselves are subject to skepticism and blame. Monster is occasionally rebuked by board posters for having a vested interest in business, for example. Interestingly, Microsoft has not attracted similar negative attention. But AARP does. Magnolia105 responds to the frustration of a fellow poster who has called for "someone with influence" to intervene in the policy arena for older workers:

> WE NEED SOMEONE ONE OUR SIDE!!! SOMEONE WITH INFLUENCE AND ANSWERS RIGHT NOW, WE PAY OUR DUES TO AARP.... THEY HAVE THE INFLUENCE, LAWYERS, AND THEMONEY. I AM NOW WONDERING WHY SOMEONE CANNOT LOBBY FOR US? AND I AM SPEAKING OF AARP....
>
> GUEST: They are the biggest lobbists in DC I've heard, only they are working on their own private agenda for their own reasons.... They are one of the richest, most affluent non-profit organizations in the world! So they are after funds for their personal issues mainly, I think.[27]

These posters showcase the insider knowledge they have about such groups as AARP in the "Question Authority" tradition for which Baby Boomers are known and for which their elders, the so-called Matures, are definitely *not* known. Magnolia105 reports that she is 62 years old. That she would daringly take on AARP in a public forum—in which AARP is the host and she the guest—runs counter to the logic we have come to expect women of her generation to exercise: Honor authority and, above all, use good manners. Some combination of her bitterness about her experience and the anonymity afforded her through the message board have perhaps liberated her from the strictures of previously imposed generational (and gendered) behavioral guidelines.[28] However, the critiques voiced by such posters as Magnolia105 and Guest are limited by their generality. They repeat that they have been told, for example, that AARP is Washington's "biggest" lobbyist, but we do not learn exactly what is meant by "biggest." They unproblematically make such claims as AARP following its own pet agenda, without noting what that agenda entails and why it is believed to be "private." So the healthy skepticism displayed by such posters is mitigated by the fuzziness with which "the facts" get reported. AARP's public agenda is no secret, however. Its Web site and other materials widely promote its lobbying on such issues as Medicare prescription drug benefits, Social Security benefits, and pension reform, for example. These are issues that clearly cut across social class lines, but they do tend to focus on retiree incomes rather than on older workers. AARP also heavily promotes travel and other "leisure" activities of the retired middle class. At one point on the message board, a poster takes AARP to task for collecting dues from people age 50 and over but not paying close attention to younger members' needs. The Web site does have a section dedicated to older workers, but it is not the site's major emphasis. As older Baby Boomers continue to increase in numbers, however, this is starting to change, as I observed between the time I started looking at the AARP Web site in early 1999 and by the time I completed this manuscript in early 2003.

 4. *References to technology in work:* Posters seeking help from fellow users "sell themselves" in terms of their technology skills, as in this post from jclyman to Thirdage.com's Job Hunting Over 50 message board:

> I am 60 years old, and absolutely healthy. I am a young thinker and not afraid to try just about anything. I have been an accountant (my degree), and in Data Processing for 19 years. I really know Windows 95, Office 97, I.E. 4.01, among other products. I was laid off (slow sales at company) last Friday.... The last time I looked for work (about 3 years ago), everyone I talked to said, "you're overqualified!"

In such posts, users almost desperately demonstrate that they have remained young through their dogged currency with technology. Generally, people who report working in information technology fields are savvier about this currency than those working in administrative office positions, as in the case of the accountant who sees it as an asset to know Windows 95.

Still, the voices of downsized IT workers in their 40s and 50s—and a few in their 60s—became a familiar refrain to me. Almost all of them worried that their field is so youth dominated that they would be unable to get work after losing their jobs. A few men in their 30s, still employed in the softening economy, wrote to the boards with similar concerns. Such workers almost universally reported that they had kept their skills sharp and their self-confidence remained high but that few people in their workplaces were as old as they were.

5. *Sharing strategies and resources:* Users frequently post Web links, most often for one of the following purposes:

   a. Breaking news on a topic of related interest, usually from a traditional news outlet, such as a major daily newspaper or network's Web site. For example, Monster.com's Jenn2 and others frequently shared developments on Congress's unemployment benefits debate via news sites.
   b. Feature stories highlighting some aspect of older workers, often unemployment or technology. When such stories complimented older workers, the poster heralded the link as accurate. When the story appeared to marginalize older workers—or stir sympathy for younger ones—the poster accompanied the link with a biting comment.
   c. Stories about related policy issues, such as immigrant workers, sometimes from alternative news sites (more often on MSN's Age Issues Forum, hosted by wristtwister).
   d. Career how-to links, usually focusing on concerns of older workers, such as strategic résumé writing.
   e. Links to stories or policy-making groups, accompanied by messages urging political activism. One poster found several message boards on which to promote her own Web site, which deals critically with workplace age discrimination. Kara Lane posts updated links with news from her site, along with encouraging messages like this one, which she posted on Thirdage.com: "Why not write your own letter to elected officials. If they get enough nasty letters, they might begin thinking that age discrimination is a real problem for them."

Queries and advice about job hunting are common. Advice frequently is unsolicited. Both sad irony and an optimistic change of pace inform some of these exchanges, when a list newcomer seeking advice from an unknown community receives confident answers from regulars. Posters who might have been mourning their own losses the day before step up to the position of "expert" when a job-hunting neophyte asks how to position a résumé to hide age while showcasing experience. The next day, when the neophyte has ended the query, a poster who leans toward self-centered venting may resurface, having been silenced during the newcomer's visit. This practice is especially prominent on Monster.com's forum, so much so that it seems that two lists (many more perhaps?) coexist in cyberspace, but not cybertime. Like a crossroads traffic light, the community's unofficial regulations yield right of way to fresh participants who express concrete desires for practical information, while venters bide their time and monitor their behavior until the light changes and they can resume at full speed.

Roger Slack and Roger A. Williams have theorized the complex interplay of Internet interfaces, discourses, and community. They suggest that users struggle to develop community discourses that are not imagined by providers.[29] The notion of community, they contend, is a contested terrain, and the intervention of technology in community fuels the disputation of where community is and how it is to be conducted. In the aging/work Internet forums, this disputation is carried out in part through the users' insistent practices of conducting more than one community in the same space, separated in time (at least through the time of new media) by the device of the message board thread, whereby dialogue is organized, collated, and archived. Through its autoregulation of traffic (apart from the hierarchical regulation by the message board regulating authority), the broader community of the forum maintains a welcoming social space for those who share an interest in the most general ways but differ significantly on particulars.

6. *Scaling down with, or versus, dignity:* Most of the posters who report finding and accepting new positions volunteer that these accompany scaled-down realities. Lower wages mean more modest living and less interesting work to many of the posters, but they are divided over whether they also mean lowered self-esteem and loss of dignity. This division occasionally sparks sharp disagreement, because individuals feel their self-worth being threatened by community members who choose different paths from their own along the unmarked highway of recovery from job loss.

The most definitive example of this dilemma I saw occurred in an exchange involving some of the regular posters to the Age Issues Forum. Jenn2, who had worked in the IT field for several years and had been laid off three times in five years, had just received an offer to manage a Cumberland Farms convenience store for $35,000 a year. She had immediately dismissed the offer as insulting and announced to the community that she would continue her search for work that was more appropriate for her. GettinSerious, who had taken a management job at Arby's for $8 an hour, took Jenn2's comments personally.

Later, when Jenn2 accepted the Cumberland Farms position after all, she quickly reported that she had negotiated her salary up to $37,000, a raise of $2,000, and vowed to continue her search for a new IT position. It was only a few weeks later that Jenn2 posted her story about having angrily quit her convenience-store job due to harsh working conditions and what she perceived as "bait and switch" tactics on her promised earnings. Although, by quitting, she had no money coming in, Jenn2 strongly expressed her conviction that the job she had taken was beneath her, and she refused to be further victimized by continuing it.

More cynical members of the message board noted that Jenn2 spent too much of her time making "useless" posts. The underlying question seemed to be, Did Jenn2, like so many of the Age Issues Forum members, really want to work, after all, or was she using the dignity issue as a shield against exertion?

7. *Intergenerational friction in the workplace:* On all the lists, the perception emerged that cliquish young co-workers dominated offices in the current environment, and this was especially so in the IT field. Older workers felt left out of the loop, turned off by what they saw as dirty and gossipy office talk. Older women, especially, perceived that younger workers tended to distance themselves from them and that younger bosses disliked them, dismissed their capabilities, and looked for excuses to fire them so they could replace them with members of their own cohort (which some reports confirmed had happened).

Conflicts with younger co-workers and bosses were always assigned to age differences and discrimination, with the posters rarely searching to assume personal responsibility. The postings always conveyed the sense that "I was here first, and they can't adapt to me." Sharpening one's own communication skills for dealing with Generation X never emerged as an option.

JRAPRIL and Kitty3033, posting to the AARP Working Options message board, joined a lengthy thread of guests and regulars who expressed exasperation about having been passed over for employment and then finding out that a young person got the job: JRAPRIL noted that he was let go by a major telecommunications firm that kept employees in their 20s in place:

> JRAPRIL: I FEEL LIKE A DINOSAUR EVEN THOUGH I COULD KICK THEIR BUTT IN A NANO SECOND. ABOUT TO LOSE MY HOME. I'M ONLY 5 MONTS AWAY FROM MY RETIREMENT MONEY. IN EMERGENCY CASE, WE SHOULD BE ABLE TO GET OUR HANDS ON OUR MONEY. I WILL NEVER MAKE IT.
>
> KITTY: . . . My experience is in a comparatively small town. We only have 2 high schools within ten mile. So, the chances are that all of those kids know each other. The agency I'm speaking of is in the vicinity of one of those high schools and mostly they only hire high school girls. So, without any hints, who do you think are ALL working in this town in ALL the office and upper pay jobs. Yes, your right. High school girls, mostly through school programs. . . . Free or low wage, knocking themselves out to please, young and decorative, employees for the office . . . .

Kitty3033 went on to say that the younger "girls" do not know the labor laws and, unlike their elders, unwittingly would allow themselves to be taken advantage of by employers, as in agreeing to work off the clock, without pay. In this and other threads on the message boards, posters concluded that such exploitative practices not only hurt employed younger (and often immigrant) workers but, by depleting the available work load, damaged unemployed elder workers' chances at being hired.

Kitty and others on the boards complained about the discrimination they felt in reporting to younger bosses:

> With age only comes the knowledge that we are not here to be someones victim. With age only can one say, "Yes, I have been here. I have gained a level of knowledge and work eithic that can ONLY be gotten through YEARS of experience, and days of learning. But with this knowledge, I also have gained the wisdom of my worth, and that no man who is IN CHARGE of another man's (ok or woman—which I am byt the way) every breathing moment.... No job wants independent thinkers today. They all want you to ONLY observe the 20 year old bosses wishes, whims and procedures that he read in a book last summer at his Business Management Class.

8. Hiding advanced age, or determination not to: This emerged, variously, as a practical necessity, a facet of self-esteem, and as a resented cultural practice. Many female posters reported dyeing their hair and dressing in contemporary (but tasteful!) clothing to mask their advancing years as a strategy for job seeking and fitting in. Many women proudly announced that they did not "look their age." Among those who reported masking the physical signs of advanced age, many wondered why they remained unsuccessful getting jobs.

Much of the discussion about hiding age concerned how to construct résumés so that years "disappeared" and how to field what posters recognized as inappropriate and illegal questions about age in job interviews. Advice varied:

a. Some posters felt that searchers should disguise their ages through a "functional résumé," which might get them in the door for an interview when a résumé that reported when they received their education would not. "Let them deal with it then," one male poster advised.
b. One poster advised a searcher to do an MBA so he could report that fresh accomplishment on a résumé to fool biased screeners.
c. "I will never hide my age—I'm proud of it," an older woman noted in another dialogue.
d. Interviewers who ask questions about age should be politely advised that the questions were illegal, some posters advised.
e. Others, more politically sensitive, suggested that more successful tactics would include laughing off the question to allow the interviewer to save face: "Ha—you and I both know that I'm old enough to do the job!"

9. Community building and shaping: Posters often vent about their feelings of helplessness, frustration, depression, and other reactions to their career losses. This theme is most frequently found on the Monster Age Issues board, and many of the users work to buoy the spirits of their comrades. Some of this work takes the form of strategic advice, but the maintenance of community is prominent in such threads. For example, a 40-year-old called shyone wrote to the board on July 8, 2002:

> Hi everyone, Well, I've been looking for work for almost 2 months now. I was working at a job I really enjoyed but needed to move back to my home state to be in my daughter's life again. Anyway I forgot how hard it is to find work. I go to the library daily to do my job search but for some reason the last few weeks made me very depressed.

Dirtman, the soil scientist, was one of several respondents: "Communities are good for sharing inspirational words. Feelings such as depression, grief, and loss can be turned into more positive feelings when shared in a group." He goes on to share several practical tips to help shyone, including entering a job search with a positive outlook.

Not all the community talk is supportive. Later in the same thread, a poster informed shyone that there is no point in updating her/his skills because, as an older worker in a tough economy, chances for success are "zero." In another case, after GettinSerious vents over her miserable fast food supervisor job, which resulted in losing unemployment benefits of approximately the same value, the mean-spirited Lunatic Fringe 7 pops off: "Can I have a burger & fries?" GettinSerious will not be shamed, however, and insists on maintaining the goodness of community. She replies: "Curly fries or homestyle?"

Commiseration is a chief function of the Age Issues community. Nm6262, writing from the West Coast, occasionally posts summaries of news stories about the so-called falling unemployment rate, rising personal income, or increased consumer confidence. "Have a great lunch," she will add bitingly, or "Let's all go shopping." Stirred into her sarcasm is an acknowledgement of the us-ness that constitutes the group, the shared desperate knowledge that those on the outside who are not suffering from the same problems apparently have not noticed the emperor without any clothes. Jennifer Tiernan has noted a similar use of the Internet by another marginalized group, female Vietnam veterans. Tiernan, drawing on the broader field of feminist new media scholarship, insists that marginalized women find a safe space through such forms for building supportive virtual communities that undermine the effects of patriarchal structures.[30] The commiserative function of the aging/work Internet forums adapts the traditionally feminine practice of gossip, inviting aging communicators of any gender to name and lash out against the forces that marginalize them.[31]

10. Floating dreams: This was by far the least common theme, but it appeared on each of the lists I examined and generally accompanied some account of frustration and desperation over shared circumstances of age discrimination. For example, Lisabarsotti of Monster.com's Age Issues Forum floated a proposal of establishing a nationwide consulting business of mature Internet professionals who would market their collective service virtually. Such proposals often got echoes of approval, but nothing followed. The dreams were not tethered to solid ideas about how to carry them out or, possibly, people had been scorched sufficiently by prior ventures so that they shied away from risk. Finally, as some participants noted, people in the community lacked resources to initiate anything proprietary. Collectively and individually, almost all the regulars who participate in these discussion groups portray a sense of wanting to fight City Hall, or Corporate America, or doing something else big, such as starting their own businesses through the Internet. But almost none of them convey a sense of savvy about how the big idea might be accomplished, including breaking the grand plan into identifiable actions. They advise newcomers without pause when the queries are small ("How do I change careers when I am working at a job I don't like?" "Take classes in the evening..."), but they often originate much larger-scale dreams that get kicked around on the list, sputter, and die. For example, a woman on one list fantasized about opening an employment agency for older workers because she could not land a job herself. "Great idea," everyone enthused, but the idea went nowhere. Another woman called repeatedly for an Internet-based IT business of Monster Age Issues regulars. The offer had no takers at the end of six months. People in the situation of needing to replace their jobs and feeling helpless in battling perceived discrimination can desperately envision "solutions" but are often ill-equipped to implement them.

Conclusion

Despite the admirable attempt to "lift up" their constituents, these message boards overall reinforce negative news about ageism in much the same way that similar subaltern communities' sites reinforce other "isms." Like the neo-Confederate cyborgs studied by Tara McPherson, these aging victims feel persecuted by a society that they perceive to power almost everyone other than themselves.[32] In these ritualized and mutually supportive spaces, group members are free to ignore their own need to interpret an increasingly complex world in order to survive in it. It is like ducking into a darkened movie theater that is showing classic films from the 1930s. Themes remain stable, because the stories are so consistent.

To be sure, there is room for both disagreement and pragmatism, within boundaries. Those who ignore the boundaries and suggest measures and interpretations that lie too far outside the realm of the consistent themes of the boards are ignored, penalized, or lost to other arenas, perhaps other communities. Those who stay in their reluctant community either continue their focus on finding a way out of their marginalized collective and individual positions, or they continue to commiserate about not being able to do so. Technology surrounds their experience: For many, it used to mean a professional identity. It has come to mean a virtual lifeline, a means to find solace and perhaps a re-entry into the professional life. For some, it sadly has become a way of recalling an identity that has passed them by.

CHAPTER 3
An American (Techno) Legend: Women, Age, and the Harley-Davidson Workplace

In American corporate culture, workers chase youth, just as people do in all other aspects of life. Dread of old age has permeated cultures for thousands of years, although how old age gets defined has been shaped and reshaped by a variety of developments, including advancing longevity. In the current economic context, however, youth rules the day in large part because of its relatively new conspiracy with digital technology in capturing the public imagination and, increasingly, the private experience. Youth, not maturity, signifies value, although disassociation from those nearing retirement has been a long entrenched organizational practice on the part of younger workers. For women who see themselves as lifetime secretaries, pursuit of knowledge about the latest technological darling for streamlining the boss-worker linkage (think Palm Pilot) is as common-sensical as plopping down five hours' pay at the cosmetic counter for a tube of "proven" wrinkle cream. What Simone de Beauvoir wrote about the fear of aging applies socially perhaps even more than physically, and it applies publicly as well as privately: "Old age looms like a calamity."[1]

Men are by no means immune to the social ravages of aging in the workplace, but women are doubly encumbered, and women of color are even further hampered.[2] In fact, in many cases, aging, for men—while younger workers may find the condition personally repugnant—may even underscore power if an executive has enough capital to begin with. Think of it as the flow chart's take on character lines. It looks somehow "natural" for the "man in charge" to be a gray-haired white guy. (The same approval is not automatically forthcoming for relatively powerless male workers, whose strengths are often judged along with perceptions of physical vigor. The boss-as-elder stereotype has also received a stern challenge in Silicon Valley, where scads of dot-com companies lack a single employee over 30 years old.) In any case, whether in board rooms or on production lines, men remain more broadly visible than women in corporate life. Women continue to occupy a narrower range of positions and, of course, earn less money. With few of them in the top ranks, the corporate struggle has hardly been won.

As women age in the corporate workplace, any claim they might have had over their job territory has been tested repeatedly. That there are increasing numbers of them due to the Baby Boom and Second Wave feminism has not seemed to matter much, especially since few pink-collar ghettos are unionized. Women in their 50s and beyond have a particularly hard time obtaining and holding on to jobs at corporations in today's labor market throughout the industrialized world.[3] The stereotype suggests that older people fail to perform well in today's escalatingly high-tech workplace, and it is more difficult for women to retain their relevance than men, who enjoy, on the whole, more empowered positions and a reputation for a more analytical nature.[4] It makes sense that women who are at midlife and somewhat older now occupy a delicate spot in corporations: They aren't yet so old that they are considered unproductive, as elderly workers unfortunately often are perceived, but they aren't young enough to have a wide array of options before them, as younger women sometimes seem to. Because of the context in which they entered the workplace (limited education, severe prejudice against women workers, and so on), most of them occupy positions with limited or no potential for advancement. What is more, this class of worker has been affected to an inordinate degree by the spiraling tendencies among corporations to "de-skill" jobs that men used to do but that women have come to perform and to "casualize" the corporate office through such practices as replacing regular employees with temporary ones, who are more affordable because they cannot claim benefits and can be gotten rid of at a moment's notice.[5] Their corporate lifeline is further attenuated by the increasing practice of "outsourcing" jobs formerly done by workers in industrialized countries to their sisters in developing countries, who are far less able to battle the corporate exploitation that comes with such low-level employment.[6]

Thirty years ago, when today's 50-year-old secretarial workers graduated from high school, they generally came to their jobs with little career ambition and knowledge of only a few office tools, including a manual typewriter and a rotary telephone. Today, they may operate a fifth-generation data-entry program or, if they have become one of the relatively few higher-paid women managers along the way, they might be trading email messages with a supplier in Japan and sharing a network calendar program.

As corporations grow in technological complexity, how are women who are now at midlife faring, and what are their perspectives on their histories and futures? This chapter answers those questions through the experience of a corporation whose image is historically about both masculinity and nonconformity—Harley-Davidson. Sixteen women who work in a range of roles at company locations in or near Milwaukee, Wisconsin, participated in long interviews and group meetings during 1998 and 1999. In their comments, they illustrated some of the tensions, inconsistencies, and victories encountered by women of advancing age who are engaged in the work of a thriving American company.

Chrome and Leather

The Spring 1999 catalog entitled "Harley-Davidson MotorClothes" was produced of heavy, semi-gloss stock. The forty-eight slick inside pages of photographs hawk sleek-cut, black-leather items that complete the rider's image-conscious wardrobe, down to the leather chaps (which feature a stretch panel that "provides ease of movement and more comfort in the saddle" for women). The message is clear that the items are expensive and the aesthete is luxuriant in a streetwise sort of way. These are not the leather things sold in shops to young, extra-hip Generation Xers. The stock includes tough, functional items for riding and trendy, expensive urban gear.

Instead of professional models, the faces of actual men and women employees of Harley-Davidson are featured modeling the attire that is for sale.[7] The first time we see an image of a "human form" is when a close-up of a snub-toed boot, framed by a blue-jeaned leg, is shown in cropped form, suggesting a rider astride a Harley. The boot, with a stainless steel buckle at the side and tire-tread sole, rests on the foot peg of the rider's Hog. Dropped in at the bottom of the photo cover is the Harley-Davidson logo, complete with its spread-winged eagle and motto: "An American legend."

Inside the catalog appear the photographs of nineteen more Harley workers, five of them young women. All of the twenty are employees at Harley-Davidson's Talladega (Alabama) Test Facility, which is pictured in a double-page spread at the beginning of the catalog. The black-and-white photo is shot at night and prominently features a tall chain-link enclosure with a top border of coiled barbwire. Tough-looking test riders, mechanics, and engineers add a graceful grittiness throughout the pages. The women, whose jobs are listed as "data entry clerks and custodians," are shunted toward the catalog's rear. Some are stereotypically clad in basic "biker-chick" attire. On the road in bike culture, a woman's role is one of support. While female consumers clearly have a more complex relationship to Harley-Davidson than this—iterations of lesbian culture, for instance, appropriate Harley symbolism quite differently—the "branding" of the motorcycles articulates the company's patriarchal ethos.

Harley-Davidson has long presented a set of cultural contradictions, wrapped around expressions of social class. After Arthur Davidson and Bill Harley built their first two-wheeled motorized vehicle in a shack in Milwaukee in 1903, the company's image emerged as a "mystique." Harley-Davidsons became symbols of male prowess and of freedom to roam. If you owned a motorcycle, you were comfortable getting grease under your fingernails. You were proudly a *man's* man.

Of course, the mystique is only partly true. Harley-Davidson's first popular motorbike, the "Silent Gray Fellow," cost $210 in 1912, for the time a substantial sum of money. During the 1950s and '60s Golden Age of Harley, when bike clubs formed around the nation, groups such as the Outlaws became known for violence and lawbreaking, and they were decidedly rough-hewn expressions of working-class, white maleness. In the late 1980s through the 1990s, the relative cost of Harley-Davidson motorcycles rose, emphasizing their value, and they grew in popularity among middle-class professional men. Harley-Davidsons became *investments*. You could buy a Harley-Davidson for $15,000 and sell it on the open market for substantially more than the purchase price. This was because demand exceeded supply, and the mystique was a commodity that seemed to carry no particular price tag. Today, according to company research, the "average" Harley rider/owner is male, 44 years old, and makes $69,000 a year as a professional. Motorcycle clubs, once famous for beer swilling and roughhousing, have largely been replaced by the Harley-Davidson–sanctioned Harley Owners Group (H.O.G.) chapters. H.O.G. chapters tend to be populated by affluent white men who had their images of maleness etched by the Harley mystique as teens and now wish to escape the stresses of business. They take their wives or girlfriends on a club ride to enjoy a weekend of fresh air and live some latter-day expression of the Easy Rider life. As in the balance of general society, H.O.G. chapters have seen a fuller participation among women, and there are certainly gay-populated chapters, but the heterosexual male-dominant mystique remains intact.

The practice of business at Harley-Davidson has also expressed this bundle of contradictions. In corporate literature, the first executive team is depicted in a photograph taken in about 1907 of the original Mr. Harley and three Brothers Davidson (four white guys wearing suits). On the last page of Harley-Davidson's 1998 annual report, the nine-member board of directors, eight white men and one somewhat younger white woman, are shown in classic corporate head shots. Harley-Davidson literature for external publics ritually connects with the workingman image, but its annual report cheerily tells shareholders of taking in $2 billion in receipts the previous year. The report isn't read by Wall Street investors only; many of the 6,000+ employees have purchased stock.

Harley-Davidson culture has long celebrated its company's foothold in American symbolism. After all, its products—hard-core American motorcycles and their accoutrement—demand a nod in this direction. For much of its history, male Harley-Davidson workers scooted from the line up the corporate ladder, and a number of the older (and decidedly male) occupants of today's office suites came from hourly jobs, not graduate programs. Many of the laborer-cum-management team members are longtime Harley-Davidson employees who sweated out the precarious market conditions of the early 1980s, surviving huge waves of layoffs on a number of "black Fridays." For quite a few men and a smaller number of women who stayed through leaner times, "sticking it out" meant moving up as market conditions improved.

For women, in corporate America generally and at Harley-Davidson, particularly, most work historically has been in administrative support structures. Only in the past couple of decades have more female assembly-line workers joined the corporation's labor pool, and they did so at a time when fewer hourly employees could penetrate the manufacturing plants' "chrome ceiling." Employees who have started out as secretaries have generally stayed in support positions, although a few female managers have emerged from the ranks. As these women get stereotyped in secretarial roles, it becomes hard for decision makers—male and female alike—to see them in any other role. Nowadays, to have a chance to climb into the driver's seat at Harley-Davidson, women and men must both arrive with one hand on a degree and the other on the throttle.

The company's parking lots make obvious the material connection between product and employee. At the manufacturing plants, the entire front façades are dedicated to parking for Harley-Davidson motorcycles owned by employees. A good number of office workers at the corporate headquarters also cycle to work. Most motorcycling commuters are men, but a growing number are women. Many employees come to work for the company already Harley-Davidson owners: They are attracted to the corporation as a workplace largely because of the attraction to the Harley mystique. Some employees come to work for the company and soon after purchase their own Harleys; a 20 percent discount not only allows employees to own their own motorcycle at an attractive price, but many buy a new bike annually.[8]

The company has an unusually progressive relationship with its unions, perhaps because of the intrinsic Harley blue-collar image. The employees, for the most part, are Harley-Davidson consumers, and many of them look and dress to fit the biker icon. The company celebrates the workers as embodiment of the mystique. Not only do the workers enjoy union-won wage and fringe benefits like other laborers, but they participate in trendy management models, such as "partnering." A management-union contract stipulates that certain decisions cannot take place in meetings without pertinent union representatives in attendance. In theory, this is the blue-collar worker's ideal: to be sitting at a table with management, making decisions that are going to affect them and their jobs, to help direct the company. This practice contributes to high morale, and in doing so has placed the company in an enviable position among its peers in manufacturing. Meetings are always attended by representatives of both labor and management, and those who attend the meetings are known in a sort of shorthand as part of the company "Leadership," not as union or management. Labor representation at such meetings often includes women and people of color, a practice Harley-Davidson has scrupulously followed, perhaps in part to atone for some unfortunate expressions of narrowmindedness by white-male managers (an occurrence that is only too common in sprawling corporations and that makes cultural critics and stockholders both wince, if for different reasons). A union representative attends all job interviews and have input in hiring, although they are far less likely to be involved in riskier strategic decisions such as those about mergers and acquisitions. At its core, then, "partnering" isn't so much a management philosophy as public relations, or an exercise in capitalism.

Participatory management, as in any corporation, hasn't meant that all workers occupy the same status. In office spaces throughout the company, employees at all levels exchange personal email messages and quietly print out children's homework projects while avoiding their supervisor's glare. The stakes are higher in such practices for lower-level employees, whose productivity is judged more constantly, and it is not unusual for nonmanagement office workers to go to great lengths to accomplish personal work in the corporate setting. Harley, in its way, is every bit as corporate as any other large company.

Harley-Davidson's technological style benefits corporation and personnel alike. The company has been at once progressive and cautious about adopting new technologies. In the 1970s, when many of the women in this study came to work at Harley-Davidson, the factory line work was dominated by the best available automation techniques of the time, and offices were supplied with IBM Selectric typewriters and cassette tapes, the written communication tools of choice. The company aggressively added tools that could help it retain its market edge and make office workers more productive, such as implementation of Computer Assisted Design programs in engineering facilities and introduction of the Wang word processors in offices. However, the company has stubbornly held on to its chief commodity, the Harley-Davidson mystique, by a prizing an aesthetic of design that proceeds from a studio model, wherein designers are first artists, then technicians. In the design rooms where some of the most important work is done, and a grandson of one of the original Mr. Davidsons is in charge, traditional drafting is prized for its purity. In each Harley-Davidson cycle, the company takes care to blend sophistication with historical integrity. Such tradition is preserved throughout Harley-Davidson, including in the tacit communication codes used by management and labor classes. Such tradition is routine for corporate America.[9] On the line in the factories, women feel a certain amount of pressure to "speak rough," as one person put it, in order to fit in; women in management jobs are informally advised to "learn to control their emotions."

Talking with Women at Harley-Davidson

Most of the information in this study came from several group meetings with three to five participants each and follow-up interviews. Group meetings proceeded with general questions, such as, "How have decisions been made about computers in your workplace?" and "How have computer systems changed since you started working?" Much discussion was dialogue among group members. Twelve women were European American, and four were of African American or other minority backgrounds. Harley-Davidson has an active affirmative action employment policy, and the company's Wisconsin workforce looks much like majority-white but multiethnic metropolitan Milwaukee. Here is a brief sketch of participants:[10]

> *Grace,* 60, an administrative support specialist who had been with the company for fifteen years. She was white, married, with a high school education, and said she used a computer at work and at home frequently. She practiced a great variety of applications.
> *Tessa,* 48, an administrative assistant, had been at Harley for twelve years and was also white, married, with a high school education. She used a computer often at work (an accounting program) and rarely at home, mostly for email.
> *Ellen,* 49, another administrative assistant, was a six-year employee who was white, divorced, and had some college but no degree. She used computers at work and at home "all the time" in a broad range of applications, including games.
> *Ingrid,* 50, was an operations director who was single and white. In her questionnaire, she did not disclose her education level. She

used computers at work and at home frequently, including game programs. Unlike the first three women, she owned her own Harley "Low Rider" and had been an avid biker for eight years.

Madeline, 51, white and divorced, was a manager in marketing and also used computers in her job and at home. She had been a Harley employee for three years and a motorcycle rider for fifteen.

Rhea, member of a non–African American minority group, a fairly highly placed manager, was divorced and had attained a master's degree. She used a variety of computer applications often at work and at home, which surprised one of the other members in the group. Rhea was very corporate in her demeanor. At 48, she had been riding a motorcycle for two years and had been at the company for twelve.

Bess, 58, had climbed the support ranks to become a coordinator in a public relations department. She had been at the company for nine years after holding other support positions elsewhere. She was white, married, with some college education, and used a large range of computer applications at work and at home. She reported that she owned and rode a Harley-Davidson but did not say for how long she had done so.

Irene, 51, had been at the company for eleven years. She was white, married, and had worked toward her master's degree. She was administrative assistant to a Harley-Davidson executive and used her computer often at work but did not own one. Her computer functions were diverse and oriented toward helping run the executive's work life. She owned a Harley-Davidson Heritage Softail Classic.

Sylvia, 53, had worked in administrative support for ten years and held a college degree. She was white, divorced, and used a computer often at work for secretarial functions but didn't want to buy one because of their quick obsolescence. She was not a rider.

Gert, who did not reveal her age, was a forklift driver, married, white, held an associate degree, and had been with the company for four years. She used computers at work and at home, mostly for schoolwork. She had been riding a Harley since she had come to the company.

Midge, 54, a fifteen-year Harley-Davidson employee, was a Human Resources manager. She was white, married, with an undergraduate degree. She used computers often at home and at work and didn't ride a motorcycle.

Bella, 53, was a twenty-four-year employee and was an executive assistant with some college education. She was white and married.

She used computers often at home and at work and didn't ride a motorcycle.

Sid, 56, was a technician who had worked at Harley for thirty years. She was a white, married, trade-school graduate who occasionally used computers at work for company email. She rode a Sportster 1200.

Gwen, was African American, had been with Harley-Davidson for twenty years and was 48 years old. She currently was working in a corporate training area of the company, but worked for a number of years on the shop floor. Gwen was married, had children, and neither she nor her husband rode a motorcycle.

Estelle, 65 years old, was also African American and had been with the company for almost thirty years. She had spent her entire career working in the manufacturing environment and currently worked second shift in an inspection area. Estelle was married and had children and grandchildren.

Rachel, 42, was African American and met with the older African American women. She had been with Harley-Davidson for only about five years. She was married with children and worked first shift on the assembly line.

Gendered Anxieties

As Juliet Webster notes, computers and related technologies are embedded in historical practices of patriarchy, and, although patriarchy and capitalism aren't the same thing, they are mutually constitutive.[11] Donna Haraway has observed that women, in traditional office roles especially, have suffered the consequences of having their work further "feminized" through microelectronics: They are made more vulnerable through technology's capabilities of rendering them more interchangeable, more exploitable than ever before.[12] In office-factory settings like Harley-Davidson, "women's work" often involves having one's services transferred from one department to another because the skill needs of units shift. Women in pink-collar positions might get moved from one physical plant location to another because of judgments about technical competency needs. In these lesser spheres of influence, women workers' memberships in placebound communities of personnel are deemed unimportant. This destabilized placement can be especially painful for older women who may have occupied departmental positions for some time and then are unceremoniously transferred to another site to provide support to people they do not know. Women who have worked together in some cases for many years now have to forge new communities and find themselves missing longtime friends who are now out of touch. This anxiety is exacerbated by fears surrounding corporate "surveillance culture." Older women I have talked with in these situations have told me of their particular fears that they will get into trouble with male management if they use company intranet systems to reach out to old friends from whom they are now separated. Younger workers are on the whole much freer in their appropriation of such technologies.

The women at Harley-Davidson expressed much interest in their relationship to computers at work as it contrasted with men's. Such contrasts seemed salient to all of them and were apparent in diverse ways. For instance, they considered themselves pragmatic in their approach to computers compared with the men they worked with or for. Such feelings ran especially high among the office women. Sylvia recalled working for an engineer who had resisted computers when they were brought into his unit ("Can you imagine?") She remembered her uneasiness over two salesmen getting brand-new machines they had no idea how to use, while she and another woman, who shared a machine, "were stuck with these junky old dinosaurs, and we were the ones that had to crank out the work." Sylvia and Bess agreed that it was a fact of corporate life that male executives got the high-tech toys, and the support staff did much of the work with much less. Several people told stories about inheriting their bosses' computers when they got new ones; the irony of better tools going to the bosses was not lost on them, and some even noted, "That's how the world works."[13]

At the production facility, Sid remembered embracing computers several years back but added that her co-workers had not unilaterally accepted them: "I think the women took it better than the men." Added Gert: "It was typing, so what was the difference for us?" Bella added:

> They hunt and peck when they type. None of the guys ever took any of those classes. Typing was women's work. And even some people feel that way today. There are guys in this office that still, they have a typewriter or they have a computer, but they still don't use them. But guys were really intimidated. I mean us girls took typing in high school.

She knew of male co-workers using their machines strategically, in a sort of competitive way, either through relating information about someone or locating information, such as whether an action had been taken as claimed. Women, Bella generalized, were unlikely to engage in such practices.

The women speculated that men experienced difficulty more often with computers than women did because "men are less patient," tolerant, and passive. Men, Bella asserted, and her fellow group members agreed, "are more interested in what a computer can do than actually doing it."

At several points, the women cited their resigned frustration with corporate structure and tended to compare the men at Harley-Davidson to their own husbands, who seemed to have similar approaches to technologies in the home. They saw technology very globally in their lives, and it would not have surprised them, I think, to hear that feminist theories and histories of technology resonated with their own more concrete views. (They did not, for the most part, however, self-identify as feminists—another matter entirely!) These women at Harley, though, with some irony, tended to view the men in charge at the corporation as almost interchangeable with the men in charge at home. They spoke of bosses and husbands who always wanted to drive, whether using office computers for purposes of maintaining control over others or going online at home to consume pleasurable entertainment like sports coverage. The women themselves felt stuck in cycles of reproduction, although at home they often felt that they more often directed themselves by choice toward the tasks involved in those cycles, because opting to use the computer had to be carved out of an already packed schedule of responsibilities. At work, they were directed to put reports together or search for information the boss could use. At home, they gravitated toward the computer to shop for home goods or make travel plans for themselves and their spouses: performing the work of leisure, the production of consumption. Reactions to these gendered uses of work and home technology ranged from resentment to acceptance. The women in their 50s and 60s felt that they had seen incremental changes toward more egalitarian contexts both at home and in the office, but they tended not to make too much of these. Boys will be boys seemed to be their underlying message.

Comparisons of computer practices in the office and at home matched Cynthia Cockburn's feminist account of gender identity and the relations of technology, which demonstrates that the introduction of technologies is always well embedded in social relations. Cockburn is concerned with how technological innovations—whether office microelectronics or the home microwave—fail to disrupt these stable gender structures: "Perhaps it is that in mainstream technology theory the key question has been how to explain change, while for feminists it seems more urgent to explain continuity."[14]

It is my belief, based on so many conversations I have had with women in their 50s, 60s, and 70s, that technological innovation does not easily destabilize existing gender relations, especially among older workers. Older women are much more likely than younger ones to find themselves in jobs where the possibility for advancement is severely limited. That plight is especially significant when we consider Ellen Seiter's observation that women drastically outnumber men in the kinds of jobs where telephones and computers are used simultaneously and that the opposite is true in more influential jobs such as computer programmer and system analyst.[15] Workers of both sexes busy themselves all day with entrenched practices that often split along gender lines. They then go home and reproduce those same sorts of practices.

Older people are much more settled in these practices than are younger workers, whose education in technology may be (though not necessarily) more democratic to begin with. In retirement, however, more optimistic pictures begin to emerge, at least in younger elders. Plenty of stereotypical uses of technology remain, with older men surfing the stock reports and playing the out-of-state lottery and their wives emailing out-of-state grandchildren and tracing genealogies. But more slippage of gendered approaches is starting to become noticeable among the newly retired folks I have met: Men picking up vacuum cleaners and emailing grandchildren, women purchasing jogging headsets and playing computer games. There are limits, of course. I have never heard an older woman talk about checking on the stock market through the Internet, and their male counterparts are unlikely to hunt for online recipes or Christmas gifts.

At Harley-Davidson, some of the women happily noted that they thought the corporate structure was growing more fair to women. Such sentiments typically were expressed by blue-collar workers, who thankfully reported on such developments as the removal of pinups from the assembly line areas as late as the mid-1990s. Other women workers did not seem to see as much progress from their seats in administrative support. Several remarked on the difficulty they had encountered through having to adapt to a computerized workplace during midlife, and some found the escalating need to adapt to change in their jobs discouraging and exhausting. Others remarked that the biggest period of change to the electronic office—the introduction of microcomputers—had come when they had faced the distraction of raising children and perhaps supporting their husbands' career efforts; now, they felt freer to concentrate on accommodating the demands of technological adaptation. These women expressed a consensus about their earlier adaptation's having more to do with everyday survival than ambition, and they believed this separated them from the men and the new crop of women.

No one romanticized the early years, including Estelle: "People always talk about the good old days. I don't want to go back to the good old days." The younger African American women listened in literal awe of her as she spoke about getting, then keeping her job at Harley-Davidson in the 1960s, when she not only had to balance the demands of wage earning and child rearing but also had to put up with the frequently expressed disrespect of some white male co-workers. In the process, she put her survival skills to work and became a union steward, effectively silencing expressions of prejudice around her. Now, as she brought her arthritis to work with her each day in anticipation of the arrival of her pension income, Estelle was quietly proud of her accomplishments in a plant whose conditions had improved demonstrably over three decades. She wore the respect of younger white, male colleagues as a badge of honor. Facing the challenge of learning about new technologies in order to do her job more efficiently in her 60s—to Harley's benefit or to her own—did not seem to Estelle to be such a large matter.

All the women conveyed a sense of emotional conflict about the measure of control the computers gave them in their work measured against the control held over them through implementation of computer systems. These concerns generally related to the women's experiences of power distribution in the workplace, but not always. Some of the frustration came from their experience of having mastered a particular software program and then coming into work one day to discover that someone had elected to change to something else. As Irene observed:

> I think a major problem with the software selection and development is that the users don't seem to get involved. I'm sure you can still find many who are still grumbling around here about our whole unhappy experiences with AmiPro, which was an absolute, total nightmare. And none of the heavy users—it might have been a fairly user-friendly package for an exec—but none of the heavy users were ever consulted or ever had a demo of the package.

Later, Irene added that she disliked wasting time having to replace personal macro files that would not convert to the program du jour: "I just feel whipped into submission." About the company's latest conversion to a Microsoft package, Irene speculated: "They'll amortize that capital investment over six years and then get something else."

Such steely business decisions do not get made exclusively by for-profit executives, however, even if in that context they do fit the logic of capital pursuit. Elementary school textbook adoptions, for example, involve similar commitments of "investment and amortization." The implication is a desire to be frugal, as in "not wasteful" with resources. But corporate frugality does not extend to all sectors of decision making, even those about technology. Sales executives, who spend the majority of their time outside the office, have plush, new, ergonomically correct chairs to sit in, while low-paid receptionists and clerks toil all day in cheaper "task chairs." In corporate America, the person who spends the most time on the golf course enjoys the priciest throne.

Harley-Davidson invests increasing increments of staff time in taking company-provided computing courses to keep up with change. Bella, at the Harley Powertrain plant, echoed secretary Irene's feelings of not being in control:

> Microsoft. AmiPro. Lotus SmartSuite. That was nothing but a glorified desktop publisher. Nobody else in the industry used that, and we were supposed to be a Fortune 500 company. It was the worst thing. People would come to me, and I could do WordPerfect in a wham-bam-thank-you-ma'am. That's what I had at

home. I used to take a lot of the AmiPro stuff home and put it in WordPerfect and do it for people when I was doing their school papers for them at work. People had a hard time here. At supplier conferences, people would actually laugh when we pulled our disks up and say, What is this? I'll be real honest. The older you are, the harder it gets. You've got to start all over again.

After twenty-four years, Bella had grown tired of relearning her job repeatedly, when the substance of work changed little. Like others, however, she appreciated the control the computers had given her over her tasks, such as when a new laptop allowed her to write up meeting minutes on the spot.

Control over work had become a thorny issue for most of these women. They at once feared and disliked "being" controlled and deeply desired the efficiency they could achieve through "having" control. Ironically, they associated both experiences of control with computers. As long as they could see themselves as controlling the machinery (instead of being controlled through it), they tended to see themselves as creative forces in the company. When they ceased to experience such feelings as creativity, they felt disempowered and unhappy.

Daily frustrations arise when a worker's computer locks up or a glitch occurs and these women must call in Harley-Davidson's technical staff. "They speak a different language than we do," Madeline said. "So that's a real disconnect. Unless you're a computer wizard, you don't know what they're talking about. The young ones don't know what they're talking about either when they start talking all these connectors and servers." Tessa observed, "Then I'm thinking, there goes another half-hour trying to get somebody to answer my question so I can print my piece of paper."

Ingrid added, "We're older. But at the same time we all have adapted to the new technology, we're all dependent on it, and we're all frustrated when it doesn't work out to our expectations. There are none of us who are saying, 'Computers are for somebody else. I'm happy with my typewriter.'" The women have cured much of their frustration by circumventing the technical staff and asking knowledgeable, accessible co-workers for help. Several mentioned that their male co-workers had joined this informal support system and felt less pressured as a result. It was clear that the women were leaders in institutionalizing this over-the-back-fence strategy and were proud of it. Midlife men in lower-level corporate positions understandably feel increasingly vulnerable to the caprice of decision makers in the so-called new economy: They are more likely to be let go than women because they make more money, and as office work becomes more feminized their skills are frequently rendered redundant. It is heartening to hear accounts of men in such positions of vulnerability joining forces with their sisters in corporate offices like those at Harley-Davidson to build unofficial backbones of support. At least at the level where such networks thrive, such developments doubly empower aging and female workers.

The issues of time and money were prominent in the discussions with Harley-Davidson employees. The women tended to see themselves as managers of both commodities on behalf of the company. As Ellen, an administrative assistant, said: "When we got our computer sixteen years ago, one of the things they stressed to us was that we would be able to do things much faster. And after we had gotten going on them for a while we looked at them and said, 'Yeah, we do things faster, but you know what? We are doing a whole lot more than we ever did before.'" As an example, Ellen noted that it seemed that technology held her creativity in a stranglehold: She felt almost as if she could no longer write without the computer.

Another headquarters employee, Ingrid, said, "Because it's so easy to make changes, we almost never finish anything." Madeline added: "You nit with it. There are so many colors and backgrounds for a presentation, so you spend an enormous amount of time with it. It used to be just black and white overheads. So now you're playing with all this other stuff."

Ingrid turned the concern from time to money: "Because you're printing, you do it once and then you print it and you make your overheads. You redo it when you change it and then you think, 'Color would really make this look nice.'" With every revision, Ingrid was well aware, production cost increases. Still, these women were aware that they routinely printed documents because they liked judging a hard copy, "the real McCoy." They apparently were less comfortable than some younger colleagues with the materiality of the electronic version of a document. In this sense, they may have practiced more wasteful routines than younger workers, and some mentioned this as a shameful shortcoming.

In terms of their desire to keep tinkering with documents until they reached a beautiful state, the women identified themselves as guilt-ridden transgressors—applying their creativity selfishly and with some degree of shame—because they had grown so attuned to bottom-line profits as the all-important goal. It seemed a bit sad that these women felt such extreme reticence about framing their creativity as a justifiable pursuit toward a desirable end product, whether of profit or of some job well done, but it was not surprising that they tended to force themselves to stifle such energies as immoral and feminine impulses. Being caught using too much color toner from the Laserjet seemed a shame akin to a child's being scolded for coloring outside the lines of a drawing: It simply shouldn't be done and, therefore, if done, had to be hidden. This is a perception that most women might recognize as all too familiar.

Most of the women expressed anxieties about the velocity of their jobs continuing to increase. Especially common was the complaint that managers had turned once reasonably paced tasks into last-minute assignments because they knew the support staff could deliver at warp speed with the aid of computers. Even so, they appreciated the trade-off of electronic cut and paste in lieu of retyping, although the task might now indeed take more time. None of the women took the position that old technologies also had a wasteful side, such as those encountered when documents had to be retyped or even reconstructed due to mistakes, misplacement, or revision. It was as if the new technologies had come to represent the contemporary danger of waste and that old ones had borne the Puritanical distinction of frugality.

A few women made note of having learned the different strategies that those people on whom they depended for information deployed for using different communication technologies on the job. Irene, the executive assistant at corporate headquarters, kept mental lists of categories such as who instantly replied to email messages but didn't check voice mail until the end of the day. She could handle communication quickly and had invented a technique for increasing her own boss's job efficiency in hidden ways, making herself prized, if not indispensable.

Most of the women commented that they generally received too much email, and people reported deleting most without reading them, because they didn't have time or they were irrelevant. Several employees reported resorting to traditional means to retrieve information that they needed for their jobs—information that was available fast via computer, but the limited number of machines in their workplace meant that access was limited. Instead, these workers used the telephone or went and looked in file drawers to find data. "I'm already working overtime," Gert explained. "How can I wait to get on a computer and work more overtime?"[16]

Bella said that employees abuse the email system—and waste others' time—by failing to keep distribution lists up to date and copying people unnecessarily:

> There's a lot of people who just do email and nothing else for communication. And these people walk out through the place, and nobody says hi to anybody, nobody smiles. Nobody does anything anymore because everything is via cc:Mail [the internal email system]. You don't even know who these people are that you have to send something to because you haven't had the time to go talk to them.

At the same time, she and some of the other women in her meeting talked about email as a savior from the isolation they can feel when they work in shop areas alone or with only one or two other people. Company email encourages a social outlet while it relieves anxieties for blue-collar women about face-to-face meetings with "upscale" office personnel at corporate headquarters. As Sid, a technician who works in a small laboratory at corporate headquarters, explained:

> You know, nobody comes up there. So I don't really see people. I don't talk to people, except [her boss]. Now, we got a new guy, but we do our job and we always have a deadline, so we are not very talkative. And I don't go very often into the cafeteria because I always look kind of grubby, and everybody is dressed up [wearing clean jeans because they work in offices]. So I don't really want to go in, you know; otherwise. They look at you kind of funny.

"Not here," a few members of the group chimed. "Juneau is different," interjected Bella. (Juneau is the nickname for the company headquarters, because it is located on Juneau Avenue.) Sid continued:

> So my jeans are dirty, I'm in my work clothes. People in the elevator, when I have to go to the fifth or sixth floor [corporate offices], they kind of look at you. So when I get cc:Mail, it's kind of nice. It's like communication with somebody else. Someone alive out there. I'm kind of isolated up there.

Sid came to work at Harley-Davidson because she has long been interested in motor works and high-performance engines. Even in this male-sphere capacity, she has retained a careful femininity in her appearance—her hairstyle, for example—and is embarrassed when she encounters "dressed up" women. Here, as in other moments of dialogue, Sid and her colleagues express some slippage between the categories of gender and social class: The blue-collar women talked about those in the office as "looking nice" in a way that seemed more socially acceptable, somehow more middle class, than appearing dirty from manual labor did, even though the laborers earned more money than many of the office workers.[17] As in their childhoods, somehow it seemed more socially acceptable for males to get dirty than for themselves.

Even as they took pride in the benefits they had won in their egalitarian blue-collar occupation, such women expressed shame about their self-presentation to other *women* in the more feminized space of the office. It was significant that when such women spoke of embarrassing or otherwise undesirable encounters with office personnel, they did not include men in these stories. They assumed a respect coming from the white-collar men, who presumably had a better appreciation for the grease under their fingernails (and, by extension, perhaps envied them a bit because of it). Using company computers allowed them a new mode of self-presentation to these other women that did not proceed from the assumption that they were automatically inferior.

As with the woman who expressed anxiety about not having access to the flood of company emails, access to computers was generally a topic of significant concern. Those who do have access to computers at manufacturing sites tire of other workers approaching them for favors. One woman said that other employees who don't have easy computer access ask her to print things out. "That's how our time gets wasted." It was clear to the people who made such remarks as this that the privilege of having a coveted computer at one's factory or workstation came at a cost to control over one's own priorities.

Access not only to computers but also to particular programs was a concern for the women at the plants. Not all workers have Internet access during working hours, and those who do are supposed to use it only for company business, such as procuring motorcycle racing or competition results to build product knowledge. At corporate headquarters, office personnel sometimes view Internet content for personal consumption. People who work in a factory setting cannot walk away from a moving assembly line to use a computer, let alone to surf the Net. In an office setting, workers are apt to enjoy more privacy that results in more surfing the Net.

Home-work

As with other recent studies on women working in large organizations, it was common to hear the Harley women speak of using computers for personal reasons. Mostly, though, "personal" talk shifted to concerns about having a home computer and how it was used. Talk about personal use of office computers centered on work for company-paid undergraduate or graduate courses, and it seemed unclear to the participants exactly where the line between personal and office worlds should be drawn with respect to education pursuits. After all, many are going to school in order to become more valuable employees; but they had some understanding of the policy that encourages doing schoolwork off the clock.[18]

The women talked of another realm in which personal and office boundaries overlapped: taking office work home to do with aid of a home computer. The women took work home for a variety of reasons. Most had trouble completing work on the job and added or replaced shift time with telecommuting. As Grace said, "I have a boss that will say, 'You need to stay home and get it done,' because the phone rings and I keep getting interrupted here. I'll stay home and get the whole thing done in two hours, and it would take me eight hours or more here."

Others reported that their computers often went down at work, and they needed to use their home systems. Some said they had a better system or a system they liked working on more at home and found it easier to get the job done there. In many cases, the women took home only certain types of projects, such as budgets. Others reported avoiding taking work home entirely, because they considered their home PCs for personal use—playing games, homework, shopping—and they didn't want to ruin that experience by ceding over the machine to work. Others took work home to do "by hand" or on a company laptop. Most had home computers, but a few didn't, because of cost, and a few avoided them because they held such a strong association with work: "I'm on it all day here. The last thing I want to do is get on it when I get home."[19]

Rhea, the upper-level manager, used her home computer to do certain shopping functions ("shopping that I consider more utilitarian than personal, such as picking out floor tile"). Because she was conscious of working long hours as a single mother, Rhea wanted to capitalize on the time-saving power of her computer so she could spend time shopping at the mall with her college-age daughter.

Rachel, an African American, said that she would like to one day own a personal computer so that she could use it to pay bills, teach her daughter computer skills, communicate with others, and work on a system for her Bible studies. However, at this point she was having trouble deciding which one to purchase and did not get much encouragement from the salesperson who was helping her make her decision.

One of the older participants, Bess, had bought a computer primarily to *save* money, in part because she had learned through her work that the Internet could do this for her. She saved $900 on her last vacation cruise, and the computer was paying her social dividends now, too, enabling her to keep in touch with newly retired friends who had relocated to Florida and play games such as Solitaire to amuse herself while her husband watched television, which she did not enjoy.

A number of women mentioned their gratitude to Harley for being a reliable provider of costly new computer goods. As Bella explained, "I really feel bad for people that have home computers and can't afford to upgrade that often as we at a big company can do. They're so good to you here. They teach you everything. A lot of companies expect you to learn on your own."

Seeing the World from Harley

The women in these groups seemed to enjoy comparing notes on their earlier experiences with computing, at Harley-Davidson and elsewhere. They laughed when they recalled early feelings of intimidation and reticence, along with the memory of once-popular software programs that have, happily for them, gone the way of the kick-starter.

Sid, a technician who had been at Harley for thirty years, started using computers in 1987 when she was a line worker. At that point, the computers were used in her area for simple input jobs, and she was glad to have them as a new tool:

> I had a computer at home, and I always like to try new things. It was a challenge. Later, we got our own computers here to show our shock dyno [a device used to measure movement or "bounce" of a motorcycle]. You have your instant graphs so you can see what it does before you have to put it in the bike and try it and ride it. You really visually couldn't see what it did before. Now you can compare one shock to the other. It makes testing much more precise. You can take a laptop out [on the test track] and do amazing work.

Other women told similar stories, revealing that they felt more empowered and saw their work as more advanced after initial contacts with computers. While they had been frustrated by incidents along the way, they realized they had the privilege of working in manufacturing and office environments at a critically exciting time.

The women contrasted their experiences with those of technologically advanced young people, both on the job and at home. On one hand, they expressed defensiveness and a sense of disconnection with the new employees who come to Harley-Davidson as video-game graduates who seem to have "no fear." On the other, they reported a sense of dedication in seeing that their own children were becoming computer competent, and they conveyed a sense of vicarious aspiration. They seemed conscious of the tension between these two impulses. For example, Bella spoke about her pride in her daughter, a computer programmer who "works contract for M&I Data Systems and gets $68 an hour take-home pay." In several other instances, women said the same sorts of things. They talked of their own efforts to provide educational advantages for their children, especially for daughters; for many, home computers were a priority in such endeavors.

However, even as Bella elaborated on her daughter's accomplishments, she and other women in her discussion group talked about their awkwardness in dealing with Generation Xer computer techies at Harley-Davidson, young people who were knowledgeable but not easy for the women to understand and appreciate. At the same time, they seemed excited to report that the information technology staff at the company had become more or less 50 percent female in recent years because women are not defering work in order to raise babies. "How many of us would like to go back to 1940 and be stay-at-home moms?" one woman asked rhetorically. It was obvious from her group's reaction that, although the women there clearly felt they needed their jobs, they also enjoyed working, and they especially liked working at Harley-Davidson. Overall, they spoke of having been treated well by the company, and they felt it had become increasingly friendly toward women. That the younger women in visible computer-related jobs occupied a relatively high status alongside the young men seemed to signal better times ahead, perhaps not for these particular women but for women more generally. Clearly, they took solace in that.

In individual cases, however, participants in these discussion groups expressed anxieties about not being able to effectively communicate with the Xers, who had seemingly cut their teeth on computers. Bess related an account of a telephone conversation she had had with a young man who was applying for a corporate sponsorship: "I told him I needed his grades, and he said he would just go in to the university computer and attach them and send them on email. It's not that it bothered me that he was doing this, but it bothered me that I didn't think this was possible. I think that way it's a bit hard."

In turn, Irene noted that these technological advances, which keep younger people ahead on the learning curve, meant that older employees feel the pressure to keep growing in their jobs: "It's something we just have to do. It's what it is. It's no longer where the boss comes in with a tape and says, 'Transcribe this, please,' or 'Would you come in and take dictation?' I did that for the first 25 years of my career with my little shorthand pad."

Irene and her co-workers explicitly stated that they had learned to adapt, and they were not ready to cede their work over to their juniors. When I asked the corporate worker group at one point if they felt they had been "passed by" in any way by the late arrival of computers during the span of their careers, the reaction was swift, negative, even somewhat defensive. Bess noted that she and the others had experienced Harley-Davidson's adaptation to a computerized workplace and thus had participated in a moment of excitement for corporate America. Rhea, who had only learned computers in the last few years, quickly stated that she hadn't felt passed by in the least. And Irene championed Harley-Davidson's provision of excellent training opportunities.

Not every older woman at Harley was so thankful. Grace had decided to retire early, at 60, because she felt caught in an inescapable trap of corporate logic: She worked in a department headed by a woman whom she felt had been promoted on the basis of knowing the "system" rather than her ability to lead. ("Politically, they needed a woman—it was good timing for her," Grace recalled.) Grace had attained a high level on the support staff scale and as a result could not easily find another available slot in the company at that level. She was leaving somewhat embittered after a disappointing try for a job in an office of the company that she felt would have been "perfect" for her, because it called for a blend of her administrative support and software-application talents and her skill at meeting the public. "It was obvious that the young man in charge took one look at me and thought, 'Whoa!' I knew he thought he had better think of something to say because he had no intention of hiring me. I wasn't decorative enough for the job. It would have to go to a much younger woman."

What would Grace do in retirement? She did not have plans beyond spending Mondays with her grandchildren to give her daughter a break. That would be fun, but she would miss working. It was a loss Grace would have to find a way to make her peace with.

Grace, at least, was leaving of her own volition, ostensibly retiring. Several women spoke in hushed tones about two older workers they had known who had sadly been forced out of their positions after failing to adapt to the changing microelectronic workplace. Speaking of one of these elders, one of the remaining office workers told of her team of staffers going to lengths to cover for the slower adapter, showing her many times how to use a simple application that they had quickly learned and "fixing" her work themselves. "She was able to attend as many computer classes as I was—Harley does a great job getting us to classes—but the way the classes work is really against the older worker," one of the co-workers told me. "She needed to just have some more one on one, and some documents to bring back for refreshers and perhaps some tutorials that may have been more grounded in the real world of Harley-Davidson's work so it would have made more sense. It can be harder to keep up when we're older."

Other workers who knew this same elder reported sadness about her awkward departure from the company. I asked a few of the women if they would put me in touch with her, but they, appropriately I think, refused. It was clear to them that she had been through enough already. In contexts like these, it is not unusual for workers in their 60s to accept retirement even if they do not entirely welcome it because they believe that the work world has changed without them. Such retirements can leave an elder rudderless, especially men. George, a 68-year-old office worker who retired from his job in Tennessee after the operation went online, put it this way: "I wasn't aiming to retire. I thought I would be ready when I retired. But I didn't want to struggle in my last years. I wanted to go out on my own terms." George had not been asked to leave, but nor had he felt included in significant work of his office. Meanwhile, the younger workers ran circles around him with the computers. He took his pension and, as boredom set in, got involved in church work. He rarely ran into anybody from the old job after a few months.

Biker Chicks: Proliferating Contradictions

Harley-Davidson is a complex social organization, where lines of tension and contradiction intersect: The mystique of rebellion and a corporate doctrine of openness and mutuality tug against the simple pursuit of capital.

Professional, skilled labor, management, and clerical "classes" coexist without entirely cohering. Office workers are "read" as middle class, and line workers are "read" as working class. Gender is more salient for older and (not always) less educated workers than it is for younger, better educated ones. Computer technologies are a strategic tool, a weapon, a trap, a trapping.

As the Harley-Davidson office experience establishes, women have multiple layers of identity, even when it seems that their roles are straightforward. The secretaries whom Juliet Webster has identified in her research as playing the role of "office wife" on some level embrace their subservience, taking pride in such accomplishments as making bosses look good through their own creativity with software or "hidden" competence with communications tools.[20] These same secretaries hunger for something more, even vicariously, for their daughters and younger colleagues. They exercise a certain consciousness about their own oppression and recognize that it has to do with their gender, age, and social class.

Even as these sixteen women occupy disparate positions relative to class status and job responsibility, they do emerge as members of a distinct group as marked by age and, to a great extent, by race or ethnicity. They are, largely, white women on the far edge of the Baby Boom (or just past it), women who have shared in the profound experience of working in paid positions at a profitable corporation that has transformed job routines through the intervention of communication and information technologies. These technologies have had a distinctively patriarchal expression, reflecting the gendered division of labor and technological design that has served as an indispensable tool of capitalism in its efforts to streamline corporate labor at the close of the twentieth century and opening of the twenty-first.[21]

These women belong to age cohorts that witnessed drastic technological implementations in the workplace and have had little or no control over them. Two decades ago, when the march of information technologies began to escalate, women in these cohorts were in their 30s and early 40s; although the context was fresh, the model had been installed. Women who used computers primarily were clerical workers, and information technologies were seen *as* their job; men who used computers were managers or professionals, and information technologies they used were seen as tools *for* the job.[22]

It was clear that the women at Harley-Davidson had come to view themselves outside of that model. Whether they were secretaries, line workers, or managers, they kept the technology at arm's length, prizing their identities as human workers and isolating computers as tools, even as inescapable traps of tools. This left me with a sense of optimism, which is not meant to belittle these women's profound frustrations with the patriarchally derived decision-making process that entraps them in their work. Their resilience and pragmatism underscore women's historical application of such strategies for management of their lives in many contexts. What is more, as they get older, they grow more skilled at making adjustments.

Converse truths compete. Experience has instructed these women in resignation. On some levels, they are simply too tired to put up a fuss and keep their noses down all the way to the payroll office. Their message resounded: "The men are going to do what they are going to do, so I'll concentrate my energies on how I can best adapt." These women have long been conscious of a firsthand knowledge of social history in the industrialized world: Sophisticated machines appear to enhance work capabilities, but their use, in fact, complicates and proliferates highly gendered routines for which women are responsible for maintaining.[23]

Women who occupy a range of relatively low-status positions at Harley-Davidson have been able to celebrate such small, feminine victories as enjoying computer access advantage over male peers who have not been trained in typing or being able to deliver connections between one's boss to other executives through "secret" knowledge about others' practices with communication technologies. But they are cautious about overplaying these tiny victories and appear to take greater satisfaction in the measurable progress exhibited by college-educated women of their children's generation, although, as we have seen, they are not entirely comfortable sharing the work space with this younger group. On the whole, these are women who are sufficiently seasoned to make the best of their situations. Women adapting, once again.

CHAPTER 4

"Granny, Go Ahead, You Won't Tear It Up": Central-City Elders Go Computing

(WITH JACQUELYN VINSON AND AMY LAUTERS[1])

The mounting media euphoria is unmistakable: America, you're not only getting older, you're getting better (at least, better than we used to think old people were supposed to be). Television commercials in an unprecedented volume extol the material virtues of aging—linked with leisure, good health, and spending power. Magazine articles comfort us with ongoing reports about the breakthrough "little blue pill" that can give back Granddad's get-up-and-go. Marketing gurus pop up in business publications with an increasingly familiar message of sheer delight, like this one, published under the headline "Targeting 50+: Mining the Wealth of an Established Generation" in the magazine *Direct Marketing:*

> Fifty represents unprecedented wealth. Yes, poverty will always affect some aging Americans. Others will just get by. But remember that the American economy has returned to world preeminence in the '90s, and the 50+ demographic will reap many of the rewards.... [W]hile these Americans often face the problem of supporting aged parents, millions of 50+ Americans are already beginning to inherit sizable estates from parents who achieved considerable financial success during the post–World War II era. What's more, the run-up of housing prices over the past 25 years—plus higher limits on profits of home sales—will create additional wealth.[2]

About the so-called Young Elders, aged 60–70, the magazine reports:

> True, this group is slowing down a bit, but having jettisoned most responsibilities and expenses, people in their 60s are increasingly taking advantage of good health to pursue hobbies and travel. Many continue to work part-time to stay active as well as maintain substantial levels of disposable income. Most enjoy doting on grandchildren with gifts and recreational travel.[3]

Now that Grandma and Grandpa have discovered retirement, these media reports imply, they have time and money to do the things they never knew they wanted to do, like take a vacation to Disney World, leaving the kids behind. And a thrilling new hobby has moved in to supplant the fading Sun Belt image of the Winnebago-dwelling snowbird: the "Surfin' Senior," to whom this quip is commonly attributed: "I want to know as much as my grandchildren!"[4]

New messages targeted to the middle class, presumably a thoroughly white audience, celebrate the joys of aging. A strong new trope of Successful Aging counterbalances the still-powerful traditional trope of old age found in American media: Absent or Undesirable Aging (usually due to dementia or other illness). Instead of advertising devoid of images of elders, we now have advertising riddled with salutes to the potency of the stereotypically bulging wallet of the middle-class elder. Much of this elder-inclusive advertising centers on high-tech markets: online trading services, health-related Web sites, personal computers. A growing impression of Web-savvy elders has come to suggest that older people who haven't joined the information-rich "haves" of society have failed to do so out of choice, or stodginess, or they just haven't been tapped as a market segment. For example, in *American Demographics:* "Yes, there are some people who want a computer and cannot afford them. But most Americans who do not own a computer simply do not want one."[5]

That segment certainly covers the principal researcher's (KR) mother, who at age 66 has ample Certificates of Deposit and pension money to live out her years in comfort, Medicare willing. But it doesn't apply to the thirty-six women that four graduate students and the professor (KR) talked with in the central city of Milwaukee between summer 1998 and spring 2000.[6] A few of them did own computers, but most did not. Almost all of them wanted to own one, if only for their grandchildren to use. And almost all of them were taking grant-provided computer-training classes at a local senior center, library, or urban league facility. Generally, the poorer ones among them were drawn to the classes out of desire to reposition themselves for the workplace, for jobs they needed to support themselves and perhaps dependent grandchildren. The ones with fewer money pressures seem to have been drawn to computer classes for self-actualization, as they may have been to a flower-arranging class. This chapter explores some issues these women raised relating to their views and experiences with computer technologies as older women, mostly African American, and largely working class.

Intersections of Age, Race, Gender, Social Class—and Work

The reception of computers among older women in the United States is easily oversimplified. Over-55s are a prime-growth segment for Internet marketers; in the summer of 2000, more than a year earlier than forecasters had expected, women overtook men as the majority of Internet users, and senior women helped to achieve that statistic.[7] These Internet-active women have been almost caricatured in the popular media as e-mail traders, genealogy scouts, and cruise bookers. An examination of sites popular among middle-class, white, elderly women (AARP.org, Seniornet.com, ThirdAge.com) will show plenty of evidence to support such generalizations. But in generalizations, experience is lost. The women of particular interest to me are those who weren't invited by the marketers hawking their cruises or the Mormons vending a rich array of European surnames. Through their genealogy website. The women who drew me near are those whose lives have been complicated by economic and social difficulties as well as old age. If such women are finding the on-ramp of the Information Superhighway, are they entering? And, I wondered, what kinds of traveling experiences might they be having there? A variety of experiences made each woman's stories distinct, but certain themes emerged.

Gender-biased workplaces, complicated by technology: It is simplistic to say either that information technologies have contributed to women's gains in the workplace or have held them back. In fact, computers and related technologies have opened up new career paths for women but have also made women's traditional workplace skills redundant and have led to disproportionate layoffs of women in the computerized workplace.[8] Computing is not a single, unified practice but is a loosely connected set of complex practices and technologies.[9] In the patriarchal workplace, therefore, older women, who likely have arrived on the job with lower levels of education and who likely have had access to information technology training generally only in clerical applications, occupy a distinct disadvantage contrasted with younger women, whose more ambitious and more recent education has prepared them for more challenging computer applications and the high pay and status that tend to accompany these.[10]

Complications of racism and ageism: Just as a number of feminists of color, including Dionne Brand, have pressed scholars to consider the conjoining of gender, race, and social class in the structures of patriarchal capitalism, we would press for inclusion of age as a significant sphere of influence. Brand claims that "black women's work" is ideologically situated, often located in the domestic-sphere equivalent of institutionalized wage earning.[11] This curse of inescapable black women's work seems especially harmful to older African Americans, whose wage-earning power likely has been limited to unskilled and a limited range of skilled labor jobs due to lack of access to education and civil rights. Because black women are disproportionately poor compared to whites, their financial outlook in the elder years, which largely derives from their early and midlife contexts, can be especially grim.[12]

Grandparents raising grandchildren: A startling increase is occurring in the number of children who are being raised alone for extended periods by grandparents in the United States. Grandparents have always provided a safety net for children when parents were unavailable, but now substantial numbers of such living arrangements are taking place, including 13.5 percent of African American children living with grandchildren (compared to 4.1 percent of whites).[13] A strong history of care giving across generations in black families has laid the foundation for this trend, which is attributable to liberalized child welfare reimbursement policies for kinship care, the ravages of the cocaine epidemic in the cities, the spiral of teen pregnancy and single parenthood, increases in incarcerated mothers, and AIDS, which is now the leading cause of death among African Americans aged 25–44. A combination of factors, including being single, living in poverty, and being African American, drastically increases the odds of being a caregiver for one's grandchildren. In many cases, such kinship care results in African American grandmothers, who often are younger than white grandmothers, sacrificing work opportunities to provide unplanned care for children.[14]

The challenges for grandparents raising grandchildren have been profound, ranging from health problems of children born to crack users to depression among both grandparent and grandchild.[15] Older black women are more likely than their white counterparts to have never married, live in poverty, have poor health, and have lower levels of education, all factors that lead them to be in greater need of seeking work in late life.[16] These pressures of custodial grandparenthood and economic difficulty have been complicated since the late 1990s by welfare reform, in which care-giving grandmothers increasingly have been forced to give up government support and look for work to support themselves and their grandchildren. In Milwaukee County, Wisconsin, the most populous metropolitan area in the nation's leading welfare-reform state, grandmothers caring for children in "skipped-generation" households have undertaken high-technology training on the advice of agencies charged with preparing them for and dispensing them in the workplace. One agency charged with carrying out such training has been the Milwaukee Urban League, located on the city's largely African

American, working-class North Side. The Urban League has been responsible for training hundreds of over-40 women for the computerized workplace, and agency personnel say it is becoming increasingly likely that their clients are responsible, in whole or in substantial part, for the economic welfare of grandchildren or great-grandchildren. Many of these women have worked at jobs sporadically through most of their lives, with interruptions often brought about by the need to care for children and other family members.[17] Most are not entitled to pensions, precisely because of such interrupted job histories, and others who receive benefits to care for the children of their children do not receive sufficient funds to get by.

The African American middle class: Lest I oversimplify—and overdramatize—the plight of older women and African American women elders in particular when it comes to the acquisition of computer knowledge and access to technology use, it is important to note that there is a substantial, growing black middle class, whose members are rapidly going online. Several of the women I found at the Washington Park Senior Center and the Milwaukee Public Library, and a handful at the Milwaukee Urban League, had middle-class sensibilities and, although a number of them were on fixed, low incomes, would have been surprised and insulted to hear someone speak of them as "working class." A few had undergraduate college degrees, and several more had at least some college background, mostly at a technical college downtown. These women had had what they thought of as careers, and they proudly spoke of working for thirty years at Ameritech the telephone company, or as a legal secretary, for instance. In most of these cases, the women had children and grandchildren who had attended college and perhaps occupied management-level positions.

Among our limited sample, these participants did not generally include the older women in their 70s or the youngest grandmothers, in their late 40s or early 50s. Most had been born in Milwaukee or had migrated there from the South as young children: They had grown up in black southern families in an urban northern environment. They had access to economic opportunities that many of our older participants had not. These women were less likely among the group to have grandchildren or great-grandchildren living with them, although a few did, usually with a divorced daughter. They tended to have smaller numbers of children and grandchildren than the poor women. All who had children had been married. They tended to have deeper knowledge—more experience—with computers, through work or PC ownership, and they tended to be more interested in taking classes for purposes of self-actualization than for positioning themselves for work. Two of the five white women we interviewed were in this middle-class group. Among them, Lenore had been a career legal secretary in both Milwaukee and Washington, D.C., and had lived in an affluent eastside suburb for twenty years. Because she had not married, she lacked some of the retirement benefits other white women her age enjoyed and had been forced to seek work in her 70s because she could not support herself on Social Security alone, at least not in affluent Shorewood.

Older Women at Work

In this section, we relate the experiences of several women we encountered who seemed representative of the group as a whole. The differences in their stories demonstrate how deeply embedded experiences of new technologies are in everyday life and support the theory that elders' encounters with new media are shaped as much by other cultural experiences as by age, just as we have found with traditional media.[18] In one important way, these dramas of contested identities put into play Donna Haraway's ideas about cyborg embodiment as the postmodern "woman of color," wherein the body fuses animal and machine and wherein "gender, race, or class consciousness is an achievement forced on us by the terrible historical experience of the contradictory social realities of patriarchy, colonialism, and capitalism."[19] The encounters with new technologies by the women elders in this research signal many postmodern disunities, and, as Haraway allows, such encounters can never be extricated from the circumstances of labor and reproduction, at least in the mechanical sense. On the whole, we found that the women in this area exercised the kind of work with computer technologies that Haraway theorizes about what she calls "cyborg writing"—"seizing the tools to mark the world that marked them as the other," "recoding communication and intelligence to subvert command and control."[20]

Ora-Lee: Taking Time for Herself in Retirement

Quite a number of the older women we interviewed (mostly in their 70s) had been born in the South and, without finishing high school, moved to the industrialized North to find decent-paying jobs. Most, but by no means all, had moved with husbands, and some of those husbands had proved to be unreliable mates. As a result, most of these women had worked hard in factories and worked even harder at home, raising several children. Commonly, these women had been driven to see that their children received better opportunities than they had been able to enjoy, but this proved difficult, and rarely did we speak with women even at the more middle-class senior center whose children had all completed college. Those children who had were sources of extreme pride. In the cases of most of the women at the senior center, they acknowledged difficulties in earning a living during their younger years, mostly involving racism and unreliable partners. Having on balance succeeded largely by themselves in raising their families, they felt, their retirement should include learning about and engaging in activities that they considered high status, such as computing. This interview segment between Jacquelyn Vinson (Jackie) and Ora-Lee, a 70-year-old African American mother of six who was born in Alabama and left high school after the tenth grade, typifies such experiences. Ora-Lee's daughter, a freelance computer analyst and high school language arts teacher, had given her a computer for her birthday, and she expressed a desire to join her grandchildren in their "computer craziness":

Ora-Lee: Three of my children [the three girls] have master's degrees and are working in corporate America. The boys, they're doing quite well, but you know how boys are, they aren't as ambitious as girls. They've got to try the world and then get involved in a lot of things.... I had some problems with them, but I didn't throw the baby out with the bathwater.

Jackie: What about your grandchildren? Do they ever help you out, do you ever do things together on the computer?

Ora-Lee: Uh-uh. Not on the computer, because they'll be playing games with Power somebody. They be doing games and jumping up and down....

Jackie: You must be pretty proud of your kids.

Ora-Lee: Yes. Other people ask me, Did you expect your children to turn out as well as they did? I say, sure, because I wasn't a teenager when I had those children. And I never wanted any children, and I had six. It's just a year difference in their ages.

Jackie: You really didn't want them, huh?

Ora-Lee: I said, Lord, I don't want any children, but help me to be the best mother that I can be. Every time one was born, that's what I would say.

Jackie: And He blessed you.

Ora-Lee: He surely did.

Jackie: Were you a single mother?

Ora-Lee: No. I had a husband who could educate any woman.

Jackie: Now, what does that mean?

Ora-Lee: That means, he's good today, he's bad tomorrow. Just like a cow, he'll give a good bucket of milk, then knock it over. So I'm well educated about men. I said, Lord, if you bless me with a job that I can take care of my children, I said I won't ask that man for anything. He did.

But Ora-Lee's "blessing" did not come without her own efforts. She vividly recalled almost being denied the opportunity to become the county hospital's first black cook in 1967, in part because of a racially motivated false arrest record, then whispered accounts of other, white cooks' complaints that she was making as much money as they did. Eventually, she became a union steward but had to fight vigorously on occasion to preserve this too-rare, dependable job so she could support her family.

In retirement, most of her friends filled their time with television, church work, looking after family, and living with economic hardship and chronic illness. Ora-Lee chose to learn about computers, purely for personal enrichment.

Ora-Lee: I want to be able to communicate with my grandchildren about different things [that they know about]. Help them find things, make Christmas cards. Keeping medical records. I do have that on the computer. I went to the county and I got my file when they had operated on me and different stuff like that. I have one of those files where you keep your medical information with procedures, medications, all that. I said this would be good to have on the computer. So I put it there.... I'm taking [a class at] MATC [the area technical college] now. It's a new advanced course with spreadsheets. I've had beginners and intermediate. I don't hardly know how to turn the computer on. But I get where I'm trying to go.
Jackie: So do you go on the Internet?
Ora-Lee [emphatically]: Noooooooooooo, no, no! They told me not to go on the Internet. My daughter said she'd do that for me.

Like more than half the women we spoke with, Ora-Lee expressed a fear of the Internet. She was sure it was full of danger to children, women, and old people. It was best not to mess with it. In more extreme cases, our interview participants told us that had no interest in computers, period, because of the lurking evils of the Internet associated with them.

Many of our participants told us that they used computers for some of the same applications Ora-Lee did, such as greeting-card making and connecting with grandchildren, in various ways. Ora-Lee was unusual in that she was learning to use a spreadsheet program with no plans to return to paid work; most of the Excel trainees we met were learning this program and others, such as Microsoft Word, to make themselves more fit for the employment market. A number of women who, like Ora-Lee, used the computer for personal enjoyment, used a word-processing program to store notes from their Bible classes at church. Many of these women were dedicated churchgoers and liked referencing their study materials through their home computer applications. We had the sense that using a computer to collect Bible notes may have elevated some of these women among their peers. Of these women, most were better educated than Ora-Lee. Like Ora-Lee, most of the women who had been unable to complete high school avoided word processing in favor of programs that did not demand such high literacy skills. Although virtually all of the women expressed an interest in their Christian religion, significantly none spoke of using the Internet to learn more about it or to communicate with others about it. The Internet had been used by only a few of the women, mostly to look up health-related information and perform other such research.

Jeanne: "On My Own and I Like It Fine"

Jeanne, 64, a white high school graduate, divorced her "undependable" husband after 40 years. She had four grown children and several grandchildren, most of whom lived locally. She provided free babysitting each weekend for the smallest grandchildren. She looked forward each year to Christmas, for which she was a "famous" cookie baker. She liked her relatively new independence, preferring it to living with a mate who had "his own problems." About the time of her divorce, though, she became disabled from a car crash and had to quit working as a secretary. After fighting her way through rehabilitation and facing high expenses, such as a $500 monthly health insurance premium, she needed a job. She came to the Washington Park Senior Center with her computer-training classes funded, like those of many other displaced homemakers, by a grant from the Interfaith Alliance for Older Adults. Additionally, she worked with a counselor at Interfaith in seeking a secretarial job. In her earlier secretarial work, she had adjusted from clunky precalculator instruments and manual typewriters to electronic calculators and basic computer programs. In her current training, she was catching up on the newest secretarial computing applications and the Internet. She felt accomplished. And bitter. Employers, she felt, discriminated against her because of her age.

Jeanne: They can get younger people, who are less qualified, to work cheaper, for one thing. I think it's a mistake on their part, because I think older people aren't going to be out getting married, they're more stable. They're not going to be out having a family. They're not going to be hung over....
Karen: Do you think you can go back to work full time if you get the chance?
Jeanne: I don't know if I'm going to be able to do something as a full-time job eight hours a day without some changes. Most of my jobs was diversified. I could get up, do this, do that for a while. I don't want to sit at a computer and do this here, and enter numbers and codes and letters in there all day long. That's what I was doing before, and I want to do something I can advance up to.... I don't want to let my skills go to waste.
Most older people don't like change. They like to stay in their comfort zone. They're afraid of the unknown out there. I have a mother, she's 88, she won't even have a cordless phone. If that desk was in the corner for 20 years, it shouldn't be moved. I think that attitude is why a lot of employers are afraid of hiring older people. You think we can't adjust? I learned to type on a manual typewriter. Believe me, I've adjusted. They just assume.... Age is only a number to me....

Starting in my mid-fifties, I would send out twenty résumés a week to companies here in Milwaukee. I would make the final three and go for an interview. And that would be that. Now, what was the final decision? Who knows? I think that's the worst thing. That's very discouraging.
Karen: How did you deal with that kind of rejection?
Jeanne: I hated it. I knew I was qualified. I just hated to get all dressed and go through the formalities. Rearranging my schedule and it was 90 degrees and the sweat pouring down and you're trying to put makeup on, and I'm a hot person, anyway. I would start out against 300 people, and I made the final cut. And thank you for applying, but we have selected somebody else. They keep you for about a month hanging.
Karen: Do you think it's wasting a resource of older people who may not have had the computer experience that younger people have?
Jeanne: Yeah. And maybe some of these people don't learn as fast as they did once. So it'll take them a little longer, but those people would be a good person once they're placed in a job. Older people won't just go to a job and say, oh well, I got a paycheck, and I'll look for an opportunity.

Jeanne expressed interest in continuing to work with computers even ten years later, provided her health held out, because she did not wish to become "illiterate." She looked forward to Interfaith's fulfillment of its promise to provide her with a free personal computer when she completed her training. She had big plans for filing her cookie recipes on it.

Jeanne had felt discriminated against because of her age since her mid-fifties, but she hadn't really grown bitter because she enjoyed having a positive outlook and she found bitterness impractical. Another white woman we interviewed, Lenore, a career legal secretary who had worked for a prestigious law firm in Washington, D.C., as well as for a large firm in Milwaukee, expressed a great deal of bitterness. She had returned to the marketplace because, unlike most college-educated white women in their late 60s, she had neither a spouse's earnings nor a pension to draw on for economic support. She lived in an affluent suburb and could no longer afford to do so. She had to have work.

In her interview, Lenore presented us with a file full of rejection letters from local law firms and companies where she had interviewed since her mid-fifties. Also in the file were glowing references from former employers, whose firms had closed. It was clear to Lenore that her age was an obstacle. ("I would make the cut and interview, and I know for a fact that a number of these places wound up hiring young women.") But Lenore had been a legal secretary before microcassette recorders replaced steno pads and before computers were an everyday part of law-office life. It was only now that she was learning word processing. It seemed realistic to Lenore to wonder if the younger women got the jobs because they were attractive, or simply because society is ageist. It did not seem reasonable to her to believe that employers were privileging younger women's sharper technological competency over her decades of loyalty and experience.

Agnes: Challenging Racism

Agnes, a 63-year-old African American woman taking grant-supported computer training classes at the senior center, had a desire to return to the workplace in order to survive financially and maintain her vitality. She owned a small home on the North Side, and her husband had been a construction worker on big jobs in the city. Two of her children had completed college, and one had a master's degree in counseling; she hoped he would become a minister, a role she had considered for herself as well. In the past several years, she had been working off and on as both a volunteer and a paid worker in basic education outreach for older adults. When we met her, she was recovering from the deaths of her husband and a grown son. She had four surviving children, some who lived locally and others as far away as Kentucky (following an army career). Her youngest child, who was now a single man in his late 20s, still lived at home. When he was 8, Agnes had gone to work as a teacher's aide in the Milwaukee Public School District (MPS), and she worked there for 21 years, quitting one year before she could become a certified teacher herself. Over the years, a number of grandchildren had lived with her, and she had cared for them while working full time. During her latter years in the system, she had taken a number of certification courses at the local university, including some computer training, and she had achieved the rank of teacher "peer." But she left embittered:

Agnes: Elementary teachers are very hard to work with. They are very, very much used to being in control. When, where, or what to do. And they forget. And when they do that, you as an aide or a peer, it becomes a bit much. And when I was working with 21-year-old teachers that tell me I don't know how to handle children, it gets to be very irritating.
Karen: Especially when you've raised five of them.
Agnes: Yeah, well, grandchildren, too, you know. That's why I left. I don't like the way they treat the children. I was in special ed. The majority, we did have one little white boy and one little Hispanic, and the rest were all African Americans, and they were all boys. And it just irritates me. When you can't change a thing, you do the Serenity Prayer and get to stop it.
Karen: Do you think the system itself is racist?
Agnes: Of course, it is! Sometimes it's more subtle, or not really realized. The way the elementary teachers have been programmed. But this is

where most people of color will say it's being perpetuated, rather than being eliminated. The racist regime. MPS, of course, doesn't have black teachers. And those who do have been manipulated, the majority of them. If you speak out, you're gone. Fear. As broke as I am, I could still be at MPS. But it is racist.
Karen: That must have been hard, to leave.
Agnes: Especially since little black boys are being treated that way. And I'm supposed to take it. They wanted me to go to a psychologist.... And it was a white woman.... And in the end, I wished I had not worked with her. And I told her I was very irritated with white women.... And then I get to thinking about all, your mind—well, is this just the nature of them to do this? You know, you think of Susan Smith, the Stuart man, and you can't help it. I'm not angry with the whole people. But I am angry because you don't have to be that way.... Racism costs a lot. I don't know, do you think it'll ever change?
Karen: You know, I just have to say I hope so.
Agnes: I think it will. In terms of the people of the upper level, economics. [As the African American middle class grows and more blacks have money], it will change.

Agnes had come to Interfaith and the senior center to take more computer classes while working to complete her bachelor's degree in education at the local university. She hoped that, whatever she wound up doing, it would result in helping to eliminate racism in society. She felt that the new inroads that opened up to her—from distance education courses to reading the classics—would provide useful tools. She had especially enjoyed a class in race, gender, and education, although she initially expressed doubts because it was being taught by a young white woman. Agnes's current work was providing basic education for older adults—literacy skills and basic math. It wasn't easy. In providing these lessons, Agnes says she also teaches these older adults something she very much believes herself: There's nothing wrong with keeping up with the (white) Joneses: "I can live on this side of town with a million dollars just as easily as you can live over there." She wanted to liberate the older adults she worked with from bad feelings about racism and low self-esteem, concerns she had also expressed for the black children she worked with.

But she admitted she felt stressed from such a high degree of personal investment: "When I finish these classes, I'm going to get away from people, though. Sit in an office where nobody will bother me." *Could she be serious?* we asked her. She laughed. "I probably would get fired for neglecting my job for running to find out what's going on down the hall." Yes, Agnes would search for a job in which she could use her new technological skills, but it would not be an office position, she admitted. "Besides, I don't know if I'll ever get up to 40 words a minute." She was more interested in working through issues with people than working through data entry. "Right now, in a class, I'm studying the relationship between the United States and Africa—education, gender, health. I'm not good on economics. I'm using the Internet to get my facts for my paper." Agnes told us she was interested in global communication and used Karen's experience as a transplanted Southerner to illustrate her concerns:

> You see, you lost that Southern accent. I never would have known it. You know why you've lost it, don't you? It's because of the global communication lines we've got now. If you notice, your CNN is in Georgia, but I don't hear anyone from Georgia. Of course that one with the real red hair, Lynn Rourke, has a little bit. But most of them have lost that Southern accent.

The differences in speech and much of the rest of culture were being erased by such forces, Agnes said, and she did not mourn this passing, although she also recognized that the experiences of African Americans and whites in the United States were quite distinct: "We don't have to all be alike, but we can all have the same things." She intended to reach as many people of color, old and young, as she somehow could and help convince them this was so. She would do it through the field of education. She wasn't sure how, yet, but she knew she would need technological skills to succeed.

Betti: Sharing Her Blessings

Like many African American elders we interviewed, Betti's entire perspective seemed heavily informed by her connection to the Gospel. As did many of her peer group, she believed that God "opens doors" and that it is up to individuals to walk through them. At 56 and a widow, she continued to contribute to her children both financially and by helping raise her grandchildren. She moonlighted as a nursing assistant and was on the lookout for a higher-paying job. She was a receptionist at the Urban League and had not been required to use a computer in her job, although she had noticed that the other receptionists did, for such applications as mass mailings. She got the chance to take free classes in computing through the Interfaith program at Washington Park Senior Center:

Betti: If an opportunity comes, I grab onto it. I say, Lord, maybe that's the way you bless me and let me know that's something I should get into. I have an opportunity to get some experience. You never know when your supervisor will say, get some training in computers. Once my superiors see I can do something else, other than just do the little duties that I have, then they'll say, well, she has proof of her schooling. Who knows? I could get promoted to something else by the fact that I can use the computer. I know my grammar, it all plays a hand in it. I'm thinking of moving up to a medical tran...

Jackie: Medical transcriptionist?

Betti: Yes. My grandchildren are happy to see me moving into this field. When they see me, if I can accomplish it, they'll say, if Grandma can do it, I can. And I can pass this knowledge on to other people, like at church. Letting them know that, if I can do it, you can. Your mind is like a rubber band. You can stretch it to the limit. And by me having faith in God first of all, I know it's something that's going to benefit me. It's something that's clean.... The kids in the family, they're exposed to it at school. But my peers as a whole, I don't think so, because black people always is the last people to catch on to something, or to learn something or get involved in something. So since we're a little bit behind other people, we're just now getting a little bit—the older, the middle-aged people that didn't have it in school.... We don't have that many people who are motivators toward such things.

Betti spoke ironically of a young neighbor who is working on her bachelor's degree and just bought a brand-new computer but doesn't know how to use it. "And I know how now, but I don't have one." Betti hoped for an improvement, both for herself and for the young black woman. "It makes me think, my oldest daughter should be getting into this." Although she was hurting for money, Betti expressed pride that she had conquered the challenge of completing early computing classes in the past year and might have a bright future for herself:

> I was beating up on myself. I was disgusted with me. It was all new to me. I was at the desk there and I, you know, wanted my hands to do certain things that the hands weren't at the place for. Then I got angry and cried. Right at my machine in the class, because I wanted to function. I just could see I was going to be left behind.

But Betti's computer teacher and her Interfaith job counselor pressed her to stick with it and gave her step-by-step instructions to catch up. She's grateful to them.

Like Betti, Florida, also 56, was divorced and had taken the classes and had even purchased her own home computer. She liked that her grandchildren came over and played on it, as they often did. Unlike Betti, she had some college education and tended to attribute her opportunities more to hard work and her own inquisitive nature than to God. This difference appeared repeatedly among more affluent, more educated African American elders and less affluent, more poorly educated ones, regardless of whether they were relatively young or much older. (Of course, older women generally were the less educated ones.)

Eradean: Urgency for Marcus

A number of the women were returning to work after losing welfare benefits provided by the Aid to Famlies with Dependent Children program. Some were custodial grandparents in lieu of daughters or sons whose lives had been lost or disrupted due to drug abuse, incarceration, AIDS, or other difficulties. A few of the women were enduring such hardship because of the twin pressures of work and raising grandchildren that they had given no thought at all to bringing computers into their lives. One woman in her 60s was raising her grandchildren after her daughter died of a crack overdose. She had neither time nor money for computers.

On the other hand, Eradean, a 60-year-old widow who now worked as an aide in an adult day care facility, had not been in school since the eleventh grade and now supported one of her four grandchildren. She expressed the wish to buy a new computer, having window-shopped for them without sufficient resources to pay. She thought that getting one might be the key to getting a better job and leading Marcus, her young grandson, to do better in school. "He is very clever, very bright," she said.

Another custodial grandmother, 66-year-old Charlene, had completed some high school and had enrolled in computer classes because she could not afford to support herself and her grandson, who was in her care. The boy took Ritalin for his Attentive Deficit Hyperactivity Disorder (ADHD) but still had problems. She hoped a computer would help him focus on his school work. Yet another custodial grandmother said she had seen her grandchildren produce school projects and personal ones, such as signs for their bedrooms, on the computer she had purchased for their use. She hoped that meant they had an outlet for their energies that would grow in significance to them as they grew older and more vulnerable to the street and peer influence.

Not all the custodial grandmothers who had considered purchasing a home computer for their grandchildren chose to do so. Ruby, a 73-year-old woman who had completed the ninth grade, feared she could not protect her grandchildren from the Internet: "There might be a pedophile. A sex man on there."

On balance, the custodial grandmothers, such as Eradean, shared an urgency about wanting to get computers into their homes to increase their grandchildren's (and, in many cases, great-grandchildren's) chances of living better than their parents had been able to do. They expressed a fear that if they could not provide these instruments of power—the specific source of which indeed seemed mysterious to them—their particular grandchild would be doomed to failure. With a quiet desperation and generally with a sad nod to their past child-rearing experiences, these women were refusing to allow that to happen.

Intersections on the Superhighway

In many ways, the economically challenged women—poor and borderline working/middle-class—we interviewed over the two-year span that this research was conducted resemble the more affluent elders we met who were learning to use computers in fancier senior centers, retirement villages, and lifelong learning facilities. Many, like Joanne, a 65-year-old white woman living on Milwaukee's working-class South Side—were tired of tedious lives trapped in caregiving for disabled spouses and parents and tired of hearing their children accuse them of "living in the dark ages." They were proud of their children's accomplishments and anxious about their own lack of knowledge of a sphere of life that had become so important to everyone, it seemed, except their immediate peer group. They wanted to be part of it all, and so they enrolled in computing classes. Some of them even bought their own "machines."

We spoke with several women who, for one reason or another, wished they had access to computers but felt they lacked the resources to gain it. It seemed strange to us after meeting so many women who were aware of public programs where they could gain skills and perhaps even access to their own PCs. Many participants in computer classes were aware of sources where they could purchase second-hand machines. But other older women we found just felt isolated from computers: They felt that new information technologies simply occupied a space they could not enter, because they lacked the money, or their "time had passed," or it was "too much trouble." Among these women, their desperation for their grandchildren and great-grandchildren's success cost them precious resources. Recognizing their poor urban school system's failure to provide such advantages to their grandchildren, they felt they individually must provide this holy grail so that these children could "catch up" as much as possible to children of the rich, white suburbs.

The array of responses on the part of the African American elders, ranging from suspicion of the technology to a desire to harness it in a quest for more definite equality, is embedded in their experiences of technologies as extensions of authority. These were, in many cases, women who fought for civil rights in the 1950s and 1960s and whose families continued to possess a disproportionately slight access to rights and privileges in the United States at the turn of the twenty-first century. The reputed presence of pedophiles on the Internet and the reputation of the Internet as a place where data is gathered about its users on the sly worked to frame computers as a sort of *panopticon*,[21] which some of the women understandably shunned from their homes. These women already held a healthy skepticism toward government or big business as a threatening (white) authority, which might in its vast, mysterious power usurp their hard-won civil rights. Inviting the Internet into their living rooms seemed like asking for trouble. In other cases, the reaction was more at home with Haraway's notion of cyborg citizen, opting to employ the tools of the white capitalist patriarchy with a subversive intent, or at least to try to win within the system. Sometimes when the women had turned away from this choice, they were left feeling bad about themselves.

Marta, a 67-year-old widow living in a mobile home on the South Side, had no children and was a disabled, retired parts assembler with two years of high school. Her ethnicity was Hispanic/Irish American. Her trailer was cheerily decorated and spotless, but she said she grew bored and depressed from staying home with her Labrador retriever almost all the time within the confines of her tiny, Igloo-shell home. She had enrolled in a computer class at the senior center but had dropped out after one semester because she became embarrassed when her arthritis-stricken hands would "just lock up" at the keyboard and she fell behind the rest of the class. But she sensed that computers were important, and she wished she could have stayed with the program. Her nephew, a campus security guard, owned a home computer, Marta told us, and he "has the Internet" on it. She believed the unit cost "thousands and thousands of dollars," but she had no idea what her nephew could do on it. She had never seen Internet access herself. She was proud of her nephew but felt she had missed out on something important. She recalled a friend urging her to join a computer class in 1969: "Marty, go for it. It's the thing of tomorrow!" She did not recall all the details but remembered going to the East Side to an interview to try for acceptance in the class. "A black man interviewed me, and he approached me for hanky panky," Marta said. She left quickly and never again explored taking a computer class until the failed attempt almost thirty years later. At that point, Marta had wanted to find a way to tether herself to the outside world in a manner that would allow her to be more productive than television could, more "in touch" with her young nephew than the telephone could. She had struck out, and this heaped more depression onto her.

Libby, a white, 68-year-old waitress we met at the senior center, and Vertna, a 72-year-old African American former custodian who also took classes at the senior center, were both divorced and had no earnings other than a small Social Security stipend. Both seemed to lurch into the computer classes out of desperation. "I am tired of being on my feet," Libby told us. "Seeing's how I gotta work, computers seem like a better route." Beyond learning the elementary applications, such as programs associated with Microsoft Office, neither had much of an idea of what could happen next, only that, if they stuck with it, they might land a "good job." They had given no thought to the possibilities that Lenore and Jeanne would later raise to us: Employers with "good jobs" might not welcome them, even if the law said they could not discriminate. Applicants with fresher résumés and recent, appropriate school training—and who might be perceived as better learners due to youth—would rule the day.

Like many other women we met in their first or second computer class, neither Libby nor Vertna would state much for the record about their plans for continuing. "We'll see about next semester," Vertna said. When we asked why she didn't have a feel for whether she would return, she only said, "I don't know. I hadn't thought much about it. We're just going to try this right now and see what happens."

The turnover in the classes at the senior center, the library, and the urban league was high. A minority of elders completed their computer coursework and earned certificates that pronounced them employable or computer competent. Some lost interest or focus. Others, like Marta, dropped out due to feelings of incompetence or embarrassment. Some, generally older, disappeared due to chronic health problems or lack of access to reliable transportation. The oldest participants, observed Amy Lauters—one of the co-authors here, who was a course instructor at the public library and who is white—were less likely to assert themselves in class in order to stay abreast of the work. Unlike the younger African American women in Amy's classes, she noted particularly, the elder African Americans hardly ever asked a question, asked her to slow down, or otherwise raised any issue that might have been read as a challenge to her (white?) authority. Amy hadn't noticed much about this until her work on the study, but in looking back she felt the African American elders, specifically, and older women, more generally but to a somewhat lesser degree, had lacked assertiveness in the classes. "They seemed uncomfortable making waves," Amy noted.

Jackie Vinson, the African American graduate student collaborator, noted an even more disturbing occurrence: When she and Amy interviewed African American women in their 70s, even when Jackie was the main interviewer, they looked at Amy, not Jackie, when they responded. Amy is almost ten years younger than Jackie! We sensed that such deferential habits presented difficulty for some of these women in computer classes, although we have yet to investigate this question.

Amy also observed, as did other computer instructors we spoke with, that the older African American women often had not linked concrete goals to taking the classes. In the case of younger elders, the classes clearly were about the employment everyone expected them to achieve or about particular activities they wished to pursue to "keep up with the grandkids." Some spoke with sheepish amusement about how their grandchildren had encouraged them to overcome their shyness toward the (expletive) machines. "I used to think that I would break it," one woman said. "My grandkids said, Granny, go ahead! You won't tear it up!" The transition had been liberating.

The youngest among the women we talked with tended to be facing economic crises brought about by the need to provide support for grandchildren through their own employment. For the elders of more advanced years, their future with computers always seemed cloudy. As Neil Postman has noted, new technologies always carry with them both benefits and deficits, "winners and losers"; he adds: "It is both puzzling and poignant that on many occasions the losers out of ignorance, have actually cheered the winners, and some still do."[22] In the case of working-class and even a large proportion of middle-class elders, theirs is often a case of "loser" with respect to the computer and other new technologies, and they badly want their survivors to emerge as "winners" even as they do not question whether their role as "loser" is a self-made choice or to what degree it might be. In the cases of many of the elders discussed here, they backed up this desire with their own money, attempting to shower the powerful technologies on their offspring when they could afford to. Such is the quest for immortality.

The sands are shifting, slowly. As the Baby Boomers slink into their 60s, of course, radical changes will occur both in what it means to be elderly and what computers mean in everyday life. For now, elders have disparate experiences with both helping and information technologies prior to the dominance of the microcomputer, and these disparate experiences inform their relationships to computers. Elders of limited education and economic means have pacemakers installed to make their hearts pump, wear hearing aids their physicians have told them to install, talk on cordless phones, and even sit in Medicare-provided lift chairs. They adopt these new technologies with very little thought, it seems sometimes, because these technologies clearly perform as prostheses—extensions of the body that arguably are needed. At the same time, more affluent elders sometimes, out of pride and fear, shun some of these technological prostheses (hearing aids) while embracing others (cordless phones) and even many more examples: electronic treadmills, blood pressure monitors, and burglar alarms.

Stigma, asserts Erving Goffman, comes from an abomination of the body that renders us abnormal and, in a way, dead.[23] Class and age complicate practices associated with perceptions of stigma and correction of conditions that it signifies. The adoption of assistive technologies and new media gadgets depends heavily on the potential adopter's social context. What kinds of value one might associate with a particular technology vary widely. The stigma of computer geek, for example, is a stereotype that is bound both by historical context and stereotypic notions about youth, maleness, and whiteness. At one time, "computer geek" might have been a majority perception of a certain social stigma, but it is variably perceived; certainly Bill Gates, the quintessential geek, has contributed to a counterstereotype that has weighed against the original stigma. Not all communities devalue the geek, to say the least.

Another majority perception of stigma is that one associated with the aged, decaying body, at least in Western culture. Wearing hearing aids and dentures and using walking sticks invites shame for many elders, who are understandably anxious about disappearing into irrelevance and unattractiveness. This presents especially vexing issues for Baby Boomers, who are beginning their reign as the most youth-absorbed generation of geriatrics to date. For Boomers, assistive technologies are beginning to be marketed with "micro" lines and fashion consciousness: the hearing aids that hear but are not seen, the no-line bifocals.

For other elders, especially the very old, even plain assistive technologies enhance the self-image, because these elders are increasingly aware of their vulnerability to institutionalization and other forms of dependency, which would strike down their autonomy. They are aware that hearing aids, dentures, and walkers give them mobility and communication competence that they otherwise would have been forced to surrender. They can see pacemakers and bypass surgeries in the same, life-saving light. In these elders' ready acceptance of assistive devices, then, they actually employ technologies in order to battle the stigma of the disobedient, declining body. Pride may indeed goeth before a fall, but preventing the fall is a matter of not only good sense but pride as well.

In the case of more economically challenged elders, especially the older ones among them, we have the distinct impression that they do not see microcomputers as prostheses at all but as mysterious black boxes, machines that might be either good or evil but that they might well break if they touch too many of the buttons. These elders do not always see a value in the computers that might help them to recover from any stigma in the way that assistive technologies such as hearing aids might. There is a curiosity coupled with a reticence, even fear, that may stem mostly from these elders' having never before been told that these machines might be for them, after all. For some elders, the fear is so strong that it dictates turning away from the machines. Otherwise, what seems to them like inevitable failure to master the applications that younger, abler people do so readily might vigorously underscore the elders' creeping *disabilities,* both mental and physical. Calling attention to one's decay, one's loss of memory, one's arthritis-gnarled digits hunched clumsily over the foreign keyboard seems too risky. It puts the elder on an unprotected stage of difference that feeds what David Mitchell and Sharon Snyder have identified as the "double bind" of disability—fascination/repulsion.[24] The elder who fears advancing disability fears the self as an embarrassing spectacle. She may also fear the personal cost brought by the self-realization of failure: It is easier to see oneself *as* young when one has not felt foolish trying to *look* young.

Some of these elders, displaying an indomitable strength and practicality, turn those fears aside and peer within the black box that is the computer, exercising a curiosity that is moving. After staring hardship in the face, in many cases, for most of their lives, they display a resilience, whether sparked by a quest for fulfillment or mere family survival, that prods them forward. For many of these women, their assimilation into a changed world has been a response to a desire not to relinquish their grasp on life. For others, it has been more a demand for immortality through the improved lot they can achieve for their progeny.

CHAPTER 5
Use It or Lose It: The Self-Programmable Elder

Both the so-called Matures and the Baby Boomers are a far cry from what Rob Shields has identified as the *joystick generation,* the youngsters who followed Generation X and have grown up with video gaming, cyberpunk, and "everyday virtuality."[1] Yet, as Shields goes on to suggest, the practices that have sprung up in what he has identified as *cyberia*—the predicament of social residency in the virtual suburbia of Internet life—pertains heavily to older people, especially in the world of work. Shields takes issue with the data that suggest, for instance, that children are more competent users of information and communication technology than are older adults because they are able to perform video-game-like procedures faster and more competently.

One example was from a study showing that 12- and 13-year-olds can outperform older surgeons in guiding a robot to virtually perform delicate tasks, such as tying sutures. Shields sharply faults such child-adult comparisons that suggest the obsolescence of techno-naïve adults:

> Children do not have to worry about lawsuits arising from video-games or simulations. Surgeons may well be more nervous, and more cautious with robotic equipment because they are aware of the implications of mishaps and the control given over to software which assumes and is calibrated for standardized, "normal" situations and bodies. Such comments, then, reveal an insidious romanticism of technology and an attitude among designers, trainers and boosters of such technologies that the ideal user is a child. Users should abandon caution and surrender the cares and concerns of adult citizens.[2]

My contact with older adults through Internet interviews has left me with the sense that they experience varying degrees of reflexivity about their encounters with virtual work. On the whole, however, I see no evidence that the Boomers and Matures who agreed to interviews felt they were growing so comfortable with the Internet and related technologies that these instruments would take on the "second nature" presence for them that they have done for the joystick generation.

Of course, these adults have experienced a radical shift through their "virtual lives." They consistently related to me their accounts of having their life experiences transformed through both their earlier adoption of new technologies and their continuing *revisioning* of the *multimodal* work world in which they found and continue to find themselves.[3] These seasoned workers have accumulated diverse feelings about their places in the virtual world of work, and this has involved many different responses to their questions about their pasts and futures in that world.

In his analysis of the transformation of society from one that is bound by the dominance of nature, including the physically limiting structures of space and time, to one that is "networked," Manuel Castells talks about the salience of the development of new technologies in implicating what we now experience as a "timeless time," a calling into question of working time, career patterns, and notions of retirement that had heretofore been subject mostly to biology:

> It seems that all living beings, including us, are biological clocks. Biological rhythms, whether individual, related to the species, or even cosmic, are essential to human life. People and societies ignore them at their peril. For millenniums, the rhythm of human life was constructed in close relationship to the rhythms of nature, generally with little bargaining power against hostile nature forces, so that it seemed reasonable to go with the flow, and to model the life-cycle in accordance with a society where most babies would die as infants, where women's reproductive power had to be used early, where youth was ephemeral, where growing elderly was such a privilege that it brought with it the respect due to a unique source of experience and wisdom, and where plagues would periodically wipe out a sizeable share of the population.[4]

Advanced societies, Castells argues, have tended to abandon notions of orderly life-cycles and adopt postmodernist constructs of life that are informed by increasing longevity rates, lengthening reproductive potential, and the all-around avoidance of death. As a result, prior conceptions about such matters as choices about whether to retire, or decisions about transferring to a new career, become part of the new and unpredictable features of everyday life. In fact, what it means to be young or old, as Castells points out, has been subject to revision and questioning under this transformation. The life-cycle itself is up for grabs:

Now, organizational, technological, and cultural developments characteristic of the new, emerging society are decisively undermining this orderly life-cycle without replacing it with an alternative sequence. I propose the hypothesis that the network society is characterized by the breaking down of the rhythms, either biological or social, associated with the notion of a life-cycle.[5]

What had been known as the "end of life" now is constantly being reworked, according to the new patterns that the new class of "Mature" retirees and Baby Boomer workers are playing out in conjunction with the discourses of power and possibility that they encounter in everyday life, especially in work. For Castells, any homogeneity that may have been assigned to old age, as a result, is stripped away. The mandatory "social death" of elders is over. As Castells concludes, "the social attributes of these distinct old ages will differ considerably, thus breaking down the relationship between the social condition and biological stage at the roots of the life cycle."[6]

Castells has termed this network-induced blurring of the life-cycle a "social arrhythmia."[7] One of the strongest implications of this new arrhythmia is that the social actors who are the first to encounter it as aging workers lack role models and other moorings for negotiating new social spaces. Along with workers of other ages, they experience, as Castells suggests, the compression, synchronicity, and asynchronicity of temporal events in ways for which there are no previously scripted rules for negotiation.

Although Castells does not have much else to say about older people living in the networked society, he talks at some length about the sorts of new abilities that society requires in its workers. The networked society puts pressure on labor to adapt to the need to work in flat organizations of power (networks) rather than traditional vertical hierarchies; requires highly educated workers; demands that workers take initiative; requires that the workforce be able to "reprogram" itself; and demands a "hypermobility" that is responsive to the new paradigm of network logic.[8]

For older workers, as I argued in Chapter 1, this means making adjustments that younger people never suspect that *they* will not have to make. When Castells writes of a workforce needing to be able to "reprogram itself," for example, he is talking about a complex dynamic of work-group orientation that is able to adjust to new technologically enabled patterns at every turn. This requires an understanding of how to function in a team as well as how to seek new forms of knowledge and skill through an elaborate system of lifelong learning (for which the worker must take personal responsibility). Such competencies are relatively new to aging workers, and Castells himself reveals an occupation with new generations of workers as he counts strategies for companies' success in the new networked environment: finding employees with "total commitment to the business project,"[9] willing to work at least 65 hours per week; locating college graduates with new skills in such fields as engineering and e-business (especially women); recruiting highly skilled immigrants; and locating entrepreneurial graduates with the capacity for innovation in entrepreneurial environments. Moreover, Castells, in expounding on the workforce's need to skillfully reprogram itself, heralds a labor pool that must be drawn from a "new type of education."[10] He writes, "Self-programmable labor requires a certain type of education, in which the stock of knowledge and information accumulated in the worker's mind can be expanded and modified throughout his or her working life." This means formalized educational induction into a lifelong learning state of mind, what Castells calls "learning how to learn," to mitigate the quickening obsolescence of knowledge.[11]

Implications for aging workers are mixed. First, the "Matures" of the workforce—those a generation older than the Baby Boomers—were not formally educated with a bias toward lifelong learning systems but rather in a system that conveyed a stable canon of knowledge. It was to be expected that layers of knowledge would be slathered on top of old knowledge, as in Thomas Kuhn's "normal science," but this is not a generation that was taught to expect rupture, revolution, or hyperreality.[12] For Baby Boomers, this is less true but is not completely false. Continuing education has long been a part of Boomer worker reality. For someone such as myself, for example, who graduated from college in the late 1970s, the information and communication technologies that revolutionized the workplace in the following decade meant that many of us went "back to school," in one or another form, in order to adjust to new environments that came along at what was for us relatively young ages. That we went on to adjust to the Internet and other revolutionary developments in the 1990s and beyond was no big news.

However, one part of Castells's formula for e-success seems to be increasingly missing from the Boomers' toolbox: commitment to total business success at any cost. These are longtime workaholics who are starting to expect to spend less, not more, time on the job. The networked society is looking to Generations X and Y to fill the gaps that Boomers and Matures simply are unwilling or unable to address through 65-hour work weeks and beyond.[13]

This does not mean that Baby Boomers and Matures are incapable, unwilling, or necessarily unsuccessful at participating in the world of what Castells has called "self-programmable labor." On the contrary, there are many "success stories" among them, both inside and outside the IT field and involving various aspects of information and communication technologies. Bill Gates, of course, is no slouch. But for Matures and Boomers, there are particular fields of struggle over the concept of self-programmability. As Castells allows, whether one is a member of self-programmable or generic labor is not dependent on the "qualities of the person" but is "the result of the lack of social and personal investment of intellectual capital in a given

Table 5.1

SELF-PROGRAMMABLE LABOR	GENERIC LABOR
Orientation: Flat networks	Orientation: Vertical hierarchies
Lifelong learner; proactive; "anyone can succeed"	"Capped" knowledge; reactive; self as potential victim
Self as responsible agent for success within interdependent group	Institution as responsible agent for success with individual constituent
Technology as unfolding; futuristic outlook	Technology as terminal skill to acquire, obstacle to clear; past outlook
Identifies concrete resources	Waits for resources to be identified
Healthy sense of self	Self-esteem, self-awareness, sense of purpose suffer from injuries
Perceives inclusive structures, finds permeable borders	Perceives exclusionary structures
Embraces change or is pragmatic toward it	Bewildered by or resentful toward change
Age as asset: accumulation of experience, interdependent with other aged workers	Age as liability: to be masked or used as point of complaint against injury
Government as benign	Government as inadequate protector

human being."[14] Generally speaking, the older the worker, the less likely that we will find a significant amount of such capital having been invested. When age conjoins marginalized categories of gender, race, and social class, it is logical to expect that such investment is even more rare.

In the thirty-five online interviews conducted with workers who have at least some exposure to information technology, I was able to apply Castells' categories of "self-programmable labor" and "generic labor" to the positions taken by the participants. Table 5.1 illustrates the emergence of particular patterns among individuals in either category. I want to stress that such "categorization" comes with the caveat of changeability: The interviews I conducted were an attempt to "take the pulse" of a group of workers at a particular time: Early 2003. The economic downturn in the United States that began with the collapse of the technology sector in March 2000 and was compounded by political events that followed (the presidential election of 2000; September 11, 2001; and the Bush administration's showdown with Iraq and, unintentionally, much of the rest of the world community) had a severe impact on American workers, including many of those involved in this research. Whether they as individuals might move between the circles of "self-programmable" and "generic" labor would, from a Foucauldian perspective, depend not only on changing events but also on changing discursive formations intended to anticipate and respond to those events. In other words, as I argued in Chapter 3, and as Foucault and others have shown, the construction of subjectivity is a process rather than an outcome.[15] It is helpful for the moment, however, to see how certain patterns of responses to external events have created proclivities toward Castells' notions of self-programmability and genericness.

Table 5.1 might seem to indicate a "better-ness" in being part of the self-programmable labor force, and, at least from my own research, it appears that workers with characteristics in the left-hand column generally are better off than those with characteristics on the right. This is true in such ways as higher incomes, apparently positive feelings about careers, and fewer reports of negative encounters over ageism in the workplace. What is troubling, though, is that those in the right-hand column, who are generally worse off, may seem to be passive receptors of their own unhappiness, but they report that they have experienced more difficulties with access to opportunity. Almost all the workers whom I am linking to this "generic" label related stories about perceived small-minded younger managers who undercut their contributions to the work team or about having worked for companies that overlooked the opportunity to prepare them for changes in practice.

Granted, it is important not to erase any sense of potential agency among such members of the "generic" labor pool. In other words, it is common for people who have contributed to their own losses not to recognize their own shortcomings. However, the sad aspect of the stories of these "generic laborers" is the equally important imperative to recognize that many workers—especially aging workers—do get overlooked, do get undertrained, and do get blamed, undercut, and treated as liabilities. According to my research, this seems especially true in fields or positions in which information and communication technologies play intensive and demanding roles. Of course, there are fewer and fewer fields and positions that are not so concerned with technology; in these, there has been a sufficient "deskilling" or "de-professionalization" as to render positions there available mostly to marginalized workers in any case.[16]

Interviews

About half of my 35 online contacts were IT workers and half were in non-IT fields but involved in using information and communication technologies (ICTs) for work. The group was almost evenly divided between men and women. Annual salaries were fairly evenly split in the $25,000–$50,000 category and the $50,000–$75,000 category. Several earned less than $25,000 per year (evenly split by men in graduate school and women who were not in school), and a few reported annual incomes of more than $100,000 (all males). Most interviewees were U.S. residents, but a few were from Ontario, Canada; Malaysia; and South Africa. Participants were between 48 and 81 years old, with most being between 48 and 60. All were either employed full time (including the 81-year-old) or wished to be.[17]

Self-Programmable Labor Perspective

Just under half the interviewees fell into the category that I considered to reflect Castells's "self-programmable labor" perspective. They were split about evenly between women and men and between IT and non-IT field workers. No striking age difference between these and the people in the "generic labor" perspective pool emerged. All of the international respondents fell into this group.[18]

The "self-programmables" could not all be described as "optimists," although members of this group were more than twice as likely to display an optimistic outlook in the interviews than those in the generic group. Among them, most had renewed their education, generally in recent years or even currently. They tended not to be naïve about what they considered to be ageist practices among managers and co-workers, but they viewed their journey through *cyberia* pragmatically. Although they tended neither to reveal bitterness nor to convey cynicism, some did share stories of "getting the shaft," but these were always followed with bootstrapping solutions. They all cited the need to adapt to new technologies and other systems, and they tended to preach about lifelong learning as a means of achieving new forms of success.

An 81-year-old resident of St. Petersburg, Florida, talked about her experience. She is a long-term "temporary" employee at a chiropractor's office, where she has "complete charge" of the office, billing, scheduling, and insurance tasks. She reported that she is satisfied with this work, including her opportunity to use her self-taught computer skills, and enjoys being treated "with respect" by people who come into the office and meet her there.

This woman acknowledged that her work life had not unfolded without difficulties due to what she saw as ageism, however. She said that she had competently performed her duties in the finance department of a large medical clinic in Indiana, where she was the oldest member of the staff:

> My supervisor was a young lady in her early 30s and made it clear to me that she had little time for the older employees. When she was mandated to cut three people, she chose me, another lady near my age, and a younger lady who was on probation. The reason given was elimination of the position, but since I had been trained to do any job in the department, I did not buy this. Because she also eliminated the younger girl, I did not have a case, but I know it was because of my age.

The woman's "recovery" through attaining her position in the chiropractor's office related in part to her feeling that her new contacts behaved respectfully toward her:

> They are pleased that the doctor has someone my age working for him. I get a lot of compliments due to the fact that I am still in the workforce and people are amazed that I have such knowledge of the computer. There is no place to go without the computer. I encourage all young people to learn as much as possible while in school and they will find that the workforce will be a much easier place. The world is wide open out there and I just regret that I have only a few more years to serve.

A 64-year-old Ontario man, with a background in electrical engineering, similarly described himself as a "success story," despite noting that much of his working life had been difficult due to the ageism he said he experienced over the past decade. He said:

> I always have used state-of-the-art technology and have coupled this with a healthy cynicism developed from my years of work experience. Some of my fellow workers consider me an "old fart." My criticism of blindly implementing every new piece of technology or software that comes along is taken as evidence of this despite the fact that I am often proved right.

This man recognized a trend that many older workers have described to me during the course of my research: that older workers sometimes feel automatically dismissed by younger people when they express reservations about anything having to do with technology. Older workers, as a result, sometimes feel pressured to adopt the "latest and greatest" software or hardware, out of dread that they will otherwise be dismissed as stodgy, out of touch, or incompetent.

As in the cases mentioned here, several of the people I interviewed acknowledged meeting bias in the workplace and finding ways around it, or "making lemons out of lemonade." For instance, a Connecticut woman who had gone from using a hand-operated system to manufacture card keys to using a computer to do the job said she felt that both managers and co-workers who were younger will not give workers over 50 a chance to keep contributing as systems become more complicated. "I trained a younger person for my job, then got laid off," she said. "But I love computers, so I went on and got a job teaching older workers how to use them. I love it and I love the people I work with."

Like other "self-programmables," this woman touted "lifelong learning" and "continuing education" options as means of maintaining relevance to the work world. "Never stop learning," one person wrote. "Use it or lose it," added another.

Heavily represented among those who displayed the sense of self-programmability were those who professed a creative impulse having to do with microcomputers in the early days of their adoption, as with this statement from an Oregon art director, who had been working with Apple computers for commercial art for two decades:

> I saw a Macintosh computer in 1984—and fell in love... It was the first computer that could do more than word processing and databases. I could write and draw and even paint on the little black and white screen and it would print out "virtually" the same as it looked on the screen. I pulled out some retirement funds and bought one, and proceeded to read/learn as much as I could. I eventually ran the local Macintosh User group—which lead to an art directors job which lead to the consulting position with the local printer. Heady days where I helped transform them from traditional to full-digital-training employees, researching and training clients. Now I teach (part-time) how graphic designers can be productive with computers... both in print production and Web.

The Oregon artist was one of an astonishing five people in my small sample of "self-programmables" who related stories about having gotten their computing start with an early Apple computer. At the time when the Apple II was released, it quickly became popular with the set of personal-computing enthusiasts who recognized a chance to do something more with desktop PCs than typesetting and database construction. That essentially one-third of the "self-programmables" in my group of fifteen heralded the Apple as their point of "geek" origination demonstrated an early concern with creativity and an interest in adaptation and flexibility to suit a changing environment.

I recall when the Apple II and the Harris pagination system came to the small Florida newsroom I ran in the mid-1980s. The young reporters in the newsroom had little need to pay attention to either instrument because they simply were to continue to file all their stories on the Harris front-end system we used. The newsroom's young artist quickly adapted to the Apple, and soon editors all over the newsroom were ordering up "informational graphics" from her to reflect the "USA Today" effect: When property was up for rezoning, the city editor ordered a locator map; when real estate sales accelerated, a graph pinpointed the trend.

Editors in their 30s, 40s, and 50s, many of whom had been along for the move from the IBM Selectric scanner-read copy system in use through the early 1980s, were now introduced to the Harris pagination system, their second generation of computer technology. The pagination machine was much less intuitive for the user than the QuarkXpress arrangements that eventually took over such newsrooms. The learning curve was steep, and our two oldest editors, each in their late 50s, tried every means of avoiding the conversion of their layout work to this system. The daily pressure cooker of moving stories and graphics, all in the desired arrangement and on tight deadlines, to a composing room typesetter was too much. One of these senior editors quickly retired and the other became the religion editor, which was, in a way, a sort of preretirement. I had the impression at the time that their age was not as important to their giving up as was their individual inflexibility and discomfort. Unlike many of the elders I have met as a researcher, these two lacked an orientation toward change.

As these slower-moving editors went the way of the dummy sheet, young Baby Boomers moved into their chairs and kept the trains running on time. Some of them developed a flair for layout that had been impossible for them to exercise when they lacked control over the technology of page composition. These Boomer editors, who had learned layout in college with traditional paper dummies, left the womb of the composing room and its Exacto-knife-wielding paste-up artists and went for broke learning what was new. The reporters they worked with were now carrying portable Tandy computers into stadium press boxes and elsewhere and dialing up the mainframe to file stories that would never have gotten in that edition of the paper with the old technology. For newsroom folks, as well as other kinds of workers across the country, it was time to reprogram. Some of the elders did so, but many did not, and more than a few Boomers walked away from the table without success as well.

Generic Labor Perspective

Among the "generics," most tended to reflect on their knowledge fields as complete, or as rendered obsolete by superceding technologies, but not everyone did so. For example, a 55-year-old man in the IT field acknowledged that he had constantly kept up with his field in terms of hardware and software platforms. However, when he began to suffer the consequences of what he saw as clear ageism—interviews not leading to jobs, and early layoffs in the down-turned economy—he was stuck. This Birmingham, Alabama, resident had turned bitter over what he saw as corporate greed and unfair practices toward experienced technology workers. He noticed more jobs going overseas to exploit cheaper labor, and he was aware of multiple jobs that had gone to less experienced workers who were far younger than he was.

This IT professional felt rudderless in a world where he perceived the agents to be corporate executives and government decision makers. "Everyone is a number," he warned, and he was looking to the government to enact effective policies to protect people such as himself.

Another man, a 46-year-old firefighter from Texas, worked in a federal "desk job" because federal guidelines limit firefighting to 35 and younger. He felt he was physically fit and technically knowledgeable about his discipline but that his "talents are being wasted" by not being in the field. He saw himself awaiting retirement at age 55 and was not looking ahead past that point, so he was not accruing new technology skills that he could use in a different work environment. The system, he felt, is against him.

A 49-year-old Pennsylvania woman who worked as an imaging technician for an electronic records storage company said she learned her skills at a technical school as a returning student with grown children. She had been excited about the chance to work in a field that would bring her some success, but she and her fellow 40-plus workers were gradually replaced with younger employees. Now she is only partially employed in a blue-collar job that does not require computer skills. She wonders why she ever went to school.

"Don't get any older," was her advice. "Loyalty, dedication and all that good stuff doesn't actually get you anywhere anymore. So don't be real concerned with loyalty to your employer because they sure won't give a good crap about you in the long run."

For this woman and for several other "generics" I encountered, the power structure seemed both opaque and impenetrable. They felt that they had been fooled, and they harbored anger and hurt over it. They felt they had been reduced to "just getting by." Some tended to blame IT as a monolithic field. "Stay out of IT," wrote a 48-year-old woman who reported having worked only nine of the past twenty-five months. "It's a great way to become homeless. Go into health care."

Learning and the New Age

That the "self-programmables" are adapting and succeeding is a wonderful trend, but I worry that their success may obscure some of the issues confronted by the "generics." The latter's lack of sustainable relevance in the workplace is easily obscured by a "blame the victim" logic, or one that is similar to the notion that "any child in America can grow up to be President of the United States." (Or at least any white boy.) I was tempted, myself, to feel generally dismissive toward the stories of these elders who had trouble reconciling their places in a changing society and their footing in "a timeless time." *Why can't you be more solution oriented?* I wanted to ask them. *Can't you see that you have to do something?* Easy for me to say, as a middle-class academic with access to all the bells and whistles that technology has to offer—and with tenure, to boot.

What can be done for these people? I worry much less for the middle-class segments of generations being educated today and in recent decades. They were socialized through education and other processes that school in some sense is forever, and that their continuing acquisition of competencies must in large part be self-directed, as Castells suggests. They have "learned how to learn," in other words. Of course, for marginal sectors of these same generations, the experience will be much the same as it is for the Baby Boomers and Matures in my study who expressed a tendency toward the generic labor perspective. How are they to reinvent themselves if they are not invested with tools for invention in the first place?

For workers who are older or who occupy other threatened positions of status, government has a responsibility for enabling change for the better. This means, among other things, encouragement of aggressive career development strategies for corporations, as argued for by Helen Dennis at the Andrus Gerontology Center.[19] In Chapter 10, I will return to the question of a policy agenda for aging workers and offer further solutions for consideration.

Advancing Society

As Castells noted, advanced societies replace the notion of "generic labor" with that of "self-programmable labor." Given this brief foray into 50-plus workers' comments in my Internet research, what I have learned lends credence to the idea that one of the most so-called advanced societies, the United States, has found itself in a transitional period between traditional ideas about generic labor and contemporary ideas about renewable workers. The current moment, influenced as it is by the dire unknowns of the economy and technology's implications in them, has formed an especially critical node on that transitional period.

The interviews I conducted were chock full of uncertainties, anxieties, and reformational sentiments regarding the experiences of older workers. They made creative and constant connections between their understandings of technology and their career paths. Many of the workers who reported feeling "stuck" or displaced tended to see technology as either someone else's tool or as a monster in itself, lurking in the deep beyond. On the other hand, those who harbored feelings of being in control of their own fates—or at least moderately so—regarded technology as a vehicle or pathway that they either had used or had resolved to use. They further saw their role as directly their own responsibility, whether or not they believed it should be. For example, some workers expressed disappointment or dissatisfaction that institutional authorities had not more aggressively brought them opportunities to adapt to new technologies. Still, failing those opportunities, these individuals recognized a need to compensate, as had some of the interviewees reported on in other chapters of this book.

Such experiences underscore the assertion that agency, or one's capacity for driving oneself through life, is always partial. Both cultural and social constraints and individual responses to one's situation inform the limits to which agency gets annunciated in a life. The people whose stories were told through my Internet interviews seemed to have a fairly clear understanding of the possibilities for agency in their lives. While it is to be expected that people might assign more or less blame or credit than they might to particular forces, exaggerating even their own roles in their overall experiences, that should not encourage a dismissal of people's capacity for self-awareness and self-programmability.

Finally, we should question whether self-programmability is always a good thing. There is something to be said for consistency. As Castells has persuasively argued, though, consistency in today's world is in short demand in the long run. Our need to perform consistently is folded into manageable projects, and the skills involved in this kind of success are frequently folded into obsolescence, replaced shortly with a more advanced skill. Advancement, therefore, is not necessarily a drive to stabilization. Advancement means that we always have our sea legs, never our land legs. When we behave as an advanced worker in old-fashioned, stable situations, where the fancy skills are not called for, such as when we try to introduce our elders to new technology, we are reminded of how very much we have grown accustomed to being at sea. It is like the reaction that sailors who had recently docked used to get when they were observed holding on to their beverages at the harbor-side inn: "How long have you been off the boat?" In short, they were still very much rolling with the swells.

To be in an advanced labor group, then, is to always experience a level of discomfort over not having accomplished the thing that has just come around the corner, or to know some sense of uncertainty or self-loss over not knowing the constancy of one's place in the labor pool. Constantly looking toward the future and learning proactively can, of course, be an empowering set of experiences, but they can also have an unsettling impact. Living in a series of luxury hotels in multiple cities might be exhilarating, but it also does not feel like home. The opportunity for self-programmability that the new work culture has brought with it might someday bring all rewards, but right now it is also taxing.

As with Shields's study of surgeons and video-game players, it is important to realize that the virtual life has not seemed like second nature to the thirty-five people in my interviews reported here. But it is significant that they belong to generations who have experienced a huge accumulation of technological advancement during their own lives. What these generations of workers have learned to do is adjust and to know that change is coming. Younger workers also have mastered the skill of necessary adjustment, but they may be sharper at doing so because nowhere in the corners of their minds do they expect things to remain the same.

CHAPTER 6

Wizards, Space Cowboys, and (of course) Sean Connery: Film Images of Aging Workers

"My word, they don't make them like *this* anymore."
"Right. It's still in pretty good shape."
—Exchange between valet and 007 when Bond arrives at the fat farm, in the movie *Never Say Never Again*.
The valet was speaking of Bond's classic car but 007's defensiveness betrayed his body image anxieties.

Elders, when they are seen in visual culture, often are the object of the gaze others and remain virtually inarticulate about their lived experience (seen and not heard), as Andrew Blaikie has observed.[1] Reasons are complicated, having much to do with industrial expectations of audience payoff, the classic Hollywood style of storytelling, and Hollywood's notorious lack of access for older creative workers. On the first point, it has been well documented that filmmakers produce movies for young (and, typically, for male) cinemagoers, both in the United States and for export markets. Older audiences do not make money for production companies, either through their theater attendance or, for the most part, through video purchases.[2] Second, regarding Hollywood's mode of storytelling, which is steeped in cultural ideals of youthful beauty and vigor, relatively few purposes may be served by older actors and characters, especially women, who are deemed uninteresting except as counterpoints to the main attractions of feature films. Notable exceptions do occur, such as *Cocoon*, *Driving Miss Daisy*, and *Fried Green Tomatoes*, but they are few and far between. (It is worth noting here that all three examples starred the late Jessica Tandy, who only rarely appeared in films until she was sufficiently beyond menopause so as not to be read as an aging sex symbol and therefore would not be potentially mistaken for "pathetic." What's Cyd Charisse up to lately, anyway?) On the last point, the limits of access to creative production for Hollywood's aging workforce, it is, again the exception that over-50s are involved in storytelling and development. On all points, the tide may start to turn as the 600-pound gorilla, Baby Boomers in all aspects of Hollywood production, start to discover that they don't like being forced out. Jack Nicholson, for example, long one of the most powerful actors in filmmaking, graced the screen in 2002 with *About Schmidt,* a different sort of coming-of-age story, about a retired man who seeks out a refreshed identity. We may soon find that older characters, older creative workers, and, indeed, older audiences are gaining clout. When Bruce Willis or James Earl Jones or Paul Newman or Clint Eastwood sleeps with a woman his own age, on camera, the presses surely will stop.

Generally, when aging bodies are portrayed in film and other visual media, negative stereotypes abound, immersed in the trope of decay and failure.[3] Less often do we encounter a vigorous aging worker, especially one who has mastered the controls of the technology in the setting. Instead, the point of the film more likely rests on the challenges the aging worker faces at the moment of decline in capacity. If the character is central to the story, which is unusual, the challenge likely will include victory for the elder at the end, but it will be in individual terms and not in social ones. Sean Connery will be redeemed at the end of the film *in spite of* his age, because he is and always has been James Bond. His aging characters do not themselves offer a revision of the cultural meaning of old age.

This chapter explores the relatively unexamined terrain of how Hollywood has portrayed older people in the milieu of work, especially as they respond to the challenges of technology. In all, I divided eighteen films among three general themes I observed in movies produced and distributed mostly since 1970. These themes are Fairy Tales, the Child Becomes the Father, and Revisionist Histories. Not surprisingly, most of the aging characters under discussion are male and white, but a few of the most interesting are not, though none is neither. This is not to say that Hollywood has completely failed to depict the life of an aging worker who is both female and nonwhite, but I assert that most of us would be hard pressed to recite examples.

In this chapter, I look broadly at the work and the technology with which aging characters are involved. In fairy tales, for example, aged characters frequently employ magic as a means of "black-boxing" their actions: We do not understand their means but their functions are clear. Similarly, other low-tech tools, such as automobiles and the machinery of war, have been linked with aged characters, and a few of those are illustrated here.

Less frequently, but important to the aim of this book, aging characters are represented meaningfully through their relationships with new media. We are accustomed to seeing film after film in which few if any elder characters exercise agency with new technology, from *The Matrix* to *Jumpin' Jack Flash*. This chapter analyzes some notable exceptions to the typical treatment of new technology as a young person's issue, even if many of the roles are supporting ones designed to undergird the efficacy of leading characters played by young marquee darlings. Taken together, some of these elder roles are progressive and productive, hinting at a revision that is under way in understanding technologies of the aging body.[4]

Fairy Tales: The Crone, the Wizard, and the Magic of Technology

Age, work, and technology often converge in Hollywood fairy tales. In these stories, the "technology" of the aged most often is magic. Technology as magic plays into a broader cinematic technique of relying on special effects to allow less powerful agents to engage more powerful ones, a means of employing psychic economies in compensation for disadvantage, as Simon During has stated.[5] Such a technique especially lends itself to the work of elder characters, whose function in fairy tales often is to bring about special circumstances outside the realm of the everyday world, as in the fairy godmother's work in *Cinderella*.

The fairy tale's durability springs from centuries of use as a vehicle for passing along a culture's moral truths to its citizens, young and old. Because of their role in cultural transmission to generations, in fairy tales worldwide elders have always had prominent roles. Formalist accounts of fairy tales of the past ascribe the roles of elders to a narrow range of possibilities, and these are almost always informed by gender.[6] Hollywood cinema's affinity with the fairy tale is an association owed to the tale's oral origins. Following Walter Benjamin, Pat Mellencamp has observed that the fairy tale's tradition of success through the passing off of a fiction as the teller's own experience creates a stake for the listener in hearing and remembering the story:

> Fairy tales depend on the everyday and on the present, which becomes magical, incalculable, and a bit irrational.... The everyday becomes a world of wonder and surprise, rather than the repetition and boredom often ascribed to women's daily lives.... [F]airy tales can have a dark side—the familiar can also be a nightmare.[7]

The economic and creative logic of the "remake" in Hollywood cinema abets the endurance of the fairy tale in this complex. Producers recognize the

Table 6.1 Fairy Tale Films

	CINDERELLA	THE SWORD IN THE STONE	STAR WARS	COCOON	DEATH BECOMES HER	HARRY POTTER AND THE SORCERER'S STONE
Release date	1950	1963	1977	1985	1992	2001
Genre	Children's fantasy	Children's fantasy	SciFi	Drama	Dark comedy	Adventure/Fantasy
Studio and director	Disney, Hamilton Luske, Wilfred Jackson, Clyde Geronimi	Disney, Wolfgang Reitherman	20th Century Fox, George Lucas	20th Century Fox, Ron Howard	Universal, Robert Zemeckis	Warner Bros., Chris Columbus
Leading elder(s)	Wicked queen/ stepmother Fairy godmother	Merlin the wizard	Obi-Wan Kenobe, Jedi master, near the end of his life. Part Merlin, part monk, part action hero.	Three husbands in a Florida assisted-living community: Ben, Art and Joe (Willford Brimley, Hume Cronyn, Don Ameche).	Pair of lifelong competitors, whose beauty has defined their success (Goldie Hawn, Meryl Streep), and the man (Bruce Willis) they fought over. He was a world-class plastic surgeon but the women reduced him to an undertaker's makeup man.	Dumbledore, head of Hogwarts Academy and wizard patriarch. Dumbledore (Richard Harris) is marked by a Merlin-like beard and garb. He teaches the use of magic as a force for good.

Gender lessons	Beauty is more than skin deep, but women should know when to step aside for the young.	A boy must attain wisdom to become a man.	The force is with men. Princesses are temperamental.	A man wants to be a boy, or at least to behave like one. The women will follow.	It's a wise man who knows that beauty is only skin deep. Women never do.	Hermione is smart, but Dumbledore knows that it's all about Harry.
Elder's work	Image maintenance (power play) through youth discrimination.	Preparing the future leader of the free world for adulthood and the throne.	Pass the life force on to the young. No one else can do it.	Find fountain of youth (or at least immortality).	Find fountain of youth.	Preparing the future wizards of the free world for adulthood.

contd.

Table 6.1 Continued

	CINDERELLA	THE SWORD IN THE STONE	STAR WARS	COCOON	DEATH BECOMES HER	HARRY POTTER AND THE SORCERER'S STONE
Technology	Mirror and black magic potion. Technology is dangerous, but beauty is always pure.	Moral-based magic tied to strength of character and wit. Technology is good when it is in the hands of a benevolent force (male) but dangerous when in the wrong hands (female).	Jedi tool: light saber. Death star space station, space cruisers and fighters, droids.	Space friends' cocoon pods and ship (white magic misunderstood by younger men). Technology is utopic.	Old technology: Witch's black magic (wisely avoided by "real man" Bruce Willis). New technology: Mannequin paint and Bondo. Technology is dystopic and does not belong in the hands of women, who will use it only for self-interest.	Moral-based magic tied to strength of character and wit. Technology is neither utopic nor dystopic: It all depends on the wizard. It is better to be a wizard than a muggle, though.
Supporting elders?		Archimedes, Merlin's owl.	Uncle Owen and Aunt Beru, American gothic figures. Grand Moff Tarkin, Imperial governor, victim of his greedy quest for power.	Alma (Jessica Tandy).	Lisle (Isabella Rossellini): Temptress, dominatrix, and keeper of the potion—71-year-old she-Devil (European, of course).	McGonagall (Maggie Smith), Dumbledore's second in command.

	Outcome	The world we live in
	Queen's evil magic is punished. Youth prevails.	Facelifts won't do you any good. The mirror never lies.
	Merlin's good (white male) magic prevails over evil (dark female) magic. Youth prevails as Merlin's young mentee assumes the throne.	Magic (technology) is good, as long as it is in the hands of a good white man.
	Obi-Wan lets Darth Vader strike him down with the light saber. His body vanishes and lives on with "the force" in young Luke, who is now a Jedi knight.	Believe in good and you will triumph over evil.
	Elders win immortality, but they must leave Eden (Florida, Earth) to attain it.	Death is to be avoided as the inevitable result of aging.
	Elders win immortality, but their bodies fall apart and they live on in desperate ugliness. Ernest (Willis) discovers "true immortality" through family and friends.	Death is to be embraced as the inevitable result of aging. Facelifts won't do you any good. The mirror never lies.
	Dumbledore's (white male) magic prevails. Youth prevails as young mentee Harry lives up to Dumbledore's expectations and earns his thanks.	Magic (technology) is good, as long as it is used for good by a strong (usually but not always) male wizard.

timelessness of the tale's stripped-down themes as Hollywood ones: Love conquers all, Good conquers evil, Justice must be restored. In the latter part of the twentieth century, Hollywood began to take greater license with the fairy tale, resulting at times in new ambiguities that were now possible because the conditions associated with postmodern life had necessitated at least a few wrinkles in the morality tale.

In this section, I will describe the treatment of aging, work, and technology (magic) in two classic fairy tale films and four contemporary fairy tales based on dynamic challenges of aging in contemporary society (see Table 6.1). At times, as with the sections to follow, my interpretation of the categories of aging, work, and technology is broad, and, when it is, I will provide an argument for this. Only the first two pictures are based on classic "fairy tales," but I am suggesting that all six films are just that: stories told to a culture using an array of simple characters to advance a moral, even if it is held in question at the end.

The Sword in the Stone

The Sword in the Stone is in many ways a classic treatment of the aged character in Hollywood fairy-tale movies of the 1930s through the 1980s. Merlin was powerful, wise, ambitious, and mysterious—a figure for children to take seriously.

Merlin represents all the goodness of old age: His long, white beard punctuates his male sagacity and his slight figure the wizenness of physical decline. Unlike most men, Merlin has the power to see into the future. Traditional representations often show elders to be obsessed with the past, but Merlin's knowledge uses the past as a fountain of wisdom that assists him in making judgments about what is to come. Merlin's magic is his work: It is his role to serve as mentor to the future king of England, and the medium by which he will perform this duty is his magic.

In the tale, young Arthur is a servant in the castle, and Merlin's role is to liberate him and to educate him in preparation for his future role as king. The wizard employs magic for both jobs. To liberate Arthur, or Wart as he is known initially, Merlin employs magic to automate the chores of dishwashing, sweeping, and dusting that Wart must perform before undertaking his lessons. This use of magic underscores the optimism of the time that framed the reception of assistive home technologies into postwar middle-class life. The dishwasher and vacuum cleaner arrived on the scene like magic to liberate Americans from the skullduggery of housework. Merlin wisely recognized such technologies as saviors of the oppressed (in this case, children). Wart's boss, the Keeper of the castle, condemns this practice because it is "black magic" and departs from the rigor of what he sees as honest human labor. Merlin, however, opposes this notion as cruel to Wart, whom he sees as a boy in the process of developing as a great man. The technology of smart tools should be readily available to someone whose ambitions are as significant as Arthur's.

To educate Arthur, Merlin uses his magic to transform the boy into various creatures who must vanquish particular kinds of adversaries. As a fish, Arthur must become a clever swimmer and escape a bigger fish. As a squirrel, he must avoid the advances of a love-struck female squirrel. Finally, as a bird, Arthur has to escape a hawk. In doing so, he encounters an evil female wizard, and Merlin risks his own life by turning himself into a tiny germ and defeating her for Arthur's sake. After Merlin has won, Arthur admonishes him for the risk, but Merlin assures the boy that the undertaking was worth it if Arthur learned a lesson. "Knowledge and wisdom are the real powers," the boy concludes.

The film heads toward resolution when Wart/Arthur accompanies the Keeper's son, Kay, to a sword tournament, where he will serve as Kay's squire. This disappoints Merlin, who believes the role is beneath Arthur, and the wizard deserts the boy. When the boys arrive at the tournament, Arthur panics because he has forgotten to bring Kay's sword. Naturally, he notices the sword in the stone, a tool that men have long tried and failed to pull from the rock because legend has it that whoever does so will become England's king. When Arthur shows up in the crowd with the sword, he is immediately questioned. He returns to the courtyard where the stone is located. After he replaces the sword in the stone, Kay, the bigger boy, tries to pull it out again but fails. Of course, Arthur is given a chance and succeeds.

Arthur is installed in the castle as king, and Merlin uses his magic to reappear at Arthur's side. The boy's lesson is complete, but the implication is that his education will continue through his mentor's magic.

Magic for good or evil is a theme that resonates with traditional critiques of media, including new media. Merlin's age, coupled with his goodness, presumably has granted him the wisdom necessary to use magic (his technology) for good, and this always triumphs over evil magic. The tale upholds the wisdom of age while making a decidedly pro-youth statement, as magic helps make way for the future generation to take its rightful place in society.

Cinderella

Unlike Merlin, the fairy godmother of Cinderella performs her magic by doing everything for Cinderella and teaches her nothing, except that she must be home on time or something bad will happen. Whereas Merlin uses his magic to "educate" Arthur about useful life skills that will enable him to become a man (and remove the sword from the stone), the fairy godmother uses her magic to enable Cinderella to become glamorous enough to enchant the prince and to protect her "real self" from being found out. For Cinderella, the magic terminates at midnight, and her fairy godmother has left it up to her to protect her reputation.

Other elders in the classic fairy-tale stories, almost all of which were retold by Disney, convey little more than dark, Western stereotypes of old women: the wicked stepmother of *Cinderella, Snow White*'s witch (another stepmother), and the flabby-armed octopus of *The Little Mermaid*. On the other hand, the benign and helpless Nanny in *101 Dalmations*, the dotty Elder Mr. Dawes in *Mary Poppins* (contrasted against the much younger and more vigorous Bert, both played by Dick Van Dyke), and the silly and demented Uncle Albert, also in *Mary Poppins*, represent the impotent and senile side of aging seen prominently elsewhere in Hollywood cinema and virtually everywhere else in Western popular culture.[8] Still another image of the aged in these classic tales is the sage, usually a father figure, whose job it often is to give the main, youthful character protectionist advice that the character never follows. The narrative always results in trouble for the protagonist, who is reunited with the sage at the end of the film, usually following a plot twist that demonstrates that the protagonist has come to adulthood and has displaced the father figure as the new symbol of potency.

A ready example of this is found in Disney's *The Little Mermaid*, one of a spate of Disney productions over the past two decades that demonstrate the endurance of the fairy tale but insist on a new, if limited, agency for women and characters of color. Neptune, Ariel's father, warns her sternly to stay away from the ocean's surface and insists that she cannot mix with humans. By the film's end, of course, Ariel has saved Neptune's life and taught him that her heterosexual (if cross-species) union with the prince trumps his old-world logic with its prohibition on mixed marriages. (Of course, she now is "passing" as a human, with her new legs, and has abandoned her family below the surface.)

In such films as these, including *Cinderella*, the female body is written as technology—a comment that Pat Mellencamp has made about many classical films of the early twentieth century.[9] The aging female body is corrupt, subject to the ravages of time but able to cheat fate through black magic, as with the evil stepmother of *Cinderella*, the octopus of *The Little Mermaid*, and so on. Technology is sorcery.

Star Wars

Obi-Wan Kenobe's (Alec Guiness's) job is to train Luke Skywalker (Mark Hamill) for his role as Jedi knight. Obi-Wan is the stuff of legend in Luke's farm region. He is the revered master of the light saber, the tool used by those who employ the force for good against the Imperials. He trained both Luke's father and Darth Vader (not revealed in this film to be one and the same), and Vader refers to him during their climactic duel as Master.

Obi-Wan lives in a hermit's cave and does not use either androids or fighter jets, but he understands their technologies perfectly and uses them to unravel difficulties. It is Obi-Wan who steals to the nerve center of the Empire ship to disable elaborate entrapment technologies that can detect Han Solo's space craft so that Luke, Han, Princess Leia, and friends can make their escape.

At several points in the film, Obi-Wan notes that he is "too old" for this level of activity, that he is "tired' and that is time for Luke to carry on. "The force will always be with you," he finally vows. When he encounters Vader, he tells the dark knight, "If you strike me down, I will become more powerful than you ever imagined." The two fight furiously with their light sabers until Luke discovers them and calls out to Obi-Wan. With a smile, Kenobe turns his head toward Luke and lets Vader kill him. His body immediately vanishes beneath his rumpled cloak, and Luke feels his presence within from that point on. Kenobe embodied the force and now Luke embodies Kenobe.

Star Wars is a fairy tale on many fronts, but the salient one here is the master's immortality through the preparation of the young student. Fortified by Kenobe (light, "father"), Luke lives on to slay the dark knight of evil.

Cocoon

A different kind of fairy tale emerged in the late twentieth century. The theme of eternal life's quest is an old one, but films such as *Cocoon* and *Death Becomes Her* brought new twists to it. In *Cocoon,* three "old farts" in a St. Petersburg, Florida, retirement community stumbled upon a fountain of youth belonging to aliens from the planet Anteron. The buddies, played by Wilford Brimley, Hume Cronyn, and Don Ameche, found secret alien cocoons in an abandoned swimming pool next door. One of the three had cancer, and the others, and their wives, had slowed down to a near crawl through life. After they recovered their vitality through the mysterious pods, they returned to youthful pursuits: break dancing, trophy bowling, and, happily, sex. The men had found the secret to lost suppleness, energy, sexuality, and athleticism. They employed the technology of this life force as the key to their own happiness and that of other elders close to them.

A couple who were friends of the others provided an unhappy contrast. Rose, the wife, ailed from Alzheimer's disease and had trouble remembering the friends' names. Bernie, the husband, stubbornly refused to enter the pool or to allow Rose to do so, because it interfered with the natural law. Ben, played by Wilford Brimley, pointed out, "The way nature has treated us lately, I don't mind cheating her a little." Bernie ultimately conceded that Ben and his friends had the right answer when Rose died and he was unable to revive her.

The kindly aliens, led by Brian Dennehy's character, invited the elders to accompany them back to their planet, where, as Ben told his grandson, "we won't get any older, we'll never get sick, and we won't ever die." The departure, as a boatload of senior citizens was lifted up into the round space craft under the full moon, appeared as if the group was being transported into heaven itself.

Alien pods containing glowing stored bodies of Dennehy's fellow visitors provided the magical life force that energized the elders. The story may seem ridiculous, but it is not much of a stretch to substitute the field of biotechnology for the alien substance. The film's point of view underscored a reticence about aging and the death that follows it. The characters maintained that, given a chance, they could still enjoy life, and they preferred to do so, some of them even desperately. Dennehy's character, Walter, and his friends were not of Earth, but they were humanitarian agents who generously shared their science with those who needed it and would otherwise die.

Death Becomes Her

Meryl Streep and Goldie Hawn, too old to play romantic leads opposite Bruce Willis in 1992, nevertheless shared the screen with him in this light fairy tale. In a bastardization of the Snow White story, a youthful Isabella Rossellini (her character a disguised 71-year-old), plays the witch, the keeper of the eternal-youth potion, who administers it duplicitously to Helen (Hawn) and Madeline (Streep) as they greedily pursue eternal feminine beauty. The "prince," Ernest, a cosmetic surgeon-turned-undertaker's makeup artist, is involved with both women and refuses to take the potion, preferring to face his own mortality naturally (like Bernie in *Cocoon*). Helen and Madeline themselves try on the evil-witch role, first when Helen steals Madeline's fiancé (Ernest) out of cruelty. Later, Madeline "turns herself into" a fat hag, bolted into an apartment full of cats, eating canned frosting with her fingers, and repeatedly watching a videotaped shot of the beautiful actress Helen being strangled in a film—wickedly calling to mind the memory of "Mirror, Mirror on the wall."

Once they accept the witch's potion, "Hel" and "Mad" soon discover the cruel paradox at their expense: They will not ever die and they will not age. Their bodies are no longer organic and must be "maintained" like a Buick or any other machinery to keep their "young" appearance. First, the technology of the elixir renders them eternally marketable as celebrity women: wrinkleless, sans liver spots, defying gravity's effects on the bosom and buttocks. Then, the technology of spray paint and Bondo must be deployed to help them keep up the charade or their bodies will literally fall apart.

In her analysis of aging "scary women" in cinema, Vivian Sobchack deals with the central figures of *Death Becomes Her*. She observes that the film interrogates the cultural practice of plastic surgery on two levels. The first is at the narrative level, when Ernest saves Madeline's sagging career through his surgical skill and enables her to become a movie star, then becomes her henpecked cuckold and declines into "plastic surgeon . . . to the stars—dead ones, that is." The second instance of plastic surgery operates at the representational level, Sobchack continues. Through the "magic" of special computer graphic effects, following her potion taking, Madeline experiences the sudden rejuvenation of her feminine body. Along with this "instantaneous" plastic surgery, Sobchack points out, the mysterious Rossellini priestess character delivers a foreboding message of repression: The women must keep the secret of the potion.[10] The magic of the film's technology translates into the magic of cosmetic surgery, which is soon called into question when we see that Madeline and Helen are inhabiting bodies that are technically dead: They may look young but are, in fact, decayed and un-alive. The "magic" of plastic surgery is rendered a cruel lie. The "work" of aging, for women, has been the work *against* aging, the film acknowledges, and that is a fool's enterprise.

Two primary morals are sprinkled liberally throughout the film:

- Helen and Madeline, lifelong enemies, must bond themselves together in mutual self-interest to perpetuate their eternal nonlife. "You paint my ass, and I'll paint yours," Helen tells Madeline.
- Men are wiser than women because they know it is better to live a natural life, crow's feet and all, than to defy mortality. The minister at Ernest's funeral in the film's conclusion notes that Ernest had discovered the secret to immortality: friends and family. Without the worry that these two witches had brought him, Ernest had been able to return to a satisfying career as a physician, regaining the strength, confidence, and vigor that Hel and Mad had stolen from him and his work.

Harry Potter and the Sorcerer's Stone

Hogwarts Academy is full of aging faculty and spirits, but it is run with a steady hand by Dumbledore (Richard Harris) and his able assistant McGonagle (Maggie Smith). This fairy tale is part of a larger book/film/marketing project, in which young sorcerer Harry Potter learns to use his magic for good and not evil. Merlin-like Dumbledore, with a foot-long white beard and lyrical baritone mentor's voice, is one of the orphan Harry's two primary father figures (the other being the gamekeeper Hagrid, played by Robbie Coltrane). It is Dumbledore's approval that Harry constantly seeks, because he admires the great wizard and knows that such approval translates into ultimate success for himself.

Dumbledore's technology is, again, magic: an enchanted magical phoenix, the mirror of desire, the sorcerer's stone, the wave of his own hand. He uses this technology quietly and with great effect. Dumbledore is quick to administer a punishment to young wizards who abuse their technology, but he is not a micromanager. As a sage, he knows not to interfere or rush to judgment. No one ever questions him, and even the arch-villain of the film, Voldemort, never directly challenges Dumbledore. The potency of his magic appears on a par with the wisdom of his years. In this gothic tale, technology is mastered. Young wizards like Harry, who are technically gifted, are rare. It is more usual that students accrue their gifts through many years of practice of the magic arts. But Dumbledore is smart enough to know that Harry's greatness will dwarf even his own. He pronounces Harry special, and that pronouncement carries weight at Hogwarts.

For the Harry Potter stories, a classic good-versus-evil struggle, old age represents nothing less than the complexity of an accumulation of experiences. Dumbledore and Voldemort seem ancient to Harry, as do the benign spirits that inhabit the Hogwarts halls. Old people, like young people, might be good or they might be bad, and Dumbledore's sagaciousness and generous encouragement make it clear which side he is on.

The Child Becomes the Father

I identified a large cluster of Hollywood films whose age/work/technology theater is devoted to turning over the reins of adulthood from a parent figure to a child figure, and both are generally male characters (see Table 6.2). Action and science/fiction genres are most often the home of this theme, and the usual suspects do turn up: Sean Connery, Morgan Freeman, Jon Voight. That is not to say that these films lack distinction from one another. Some are more hopeful than others about aging, about technology, and about life generally. Of the films that I have studied for this section, most are from the 1980s, when the theme was most urgent, perhaps being associated with the coming of age of the late Baby Boomers.

Never Say Never Again

A James Bond thriller might seem an odd choice for classification under this theme, but *Never Say Never Again* carried the special burden of making light of an aging Sean Connery as 007. In no small part, it was the intertextuality that surrounded this film that caused it, for me, to fit "the son becomes the father" theme. In 1983, the year the movie was released, another James Bond picture came out, Roger Moore's sixth turn as 007 in *Octopussy*. The Moore role was unremarkable and expected. What made the Connery return interesting was that he had earlier sworn off it. He returned to the role at age 53,

Table 6.2 Child Becomes the Father Films

	NEVER SAY NEVER AGAIN	WAR GAMES	BACK TO THE FUTURE	INDIANA JONES AND THE LAST CRUSADE	MISSION IMPOSSIBLE	X-MEN
Release date	1983	1983	1984	1989	1996	2000
Genre	Action	SciFi/Thriller	SciFi	Action	Action/Thriller	SciFi/Action
Studio and director	Warner Bros., Irvin Kershner	MGM, John Badham	Universal, Robert Zemeckis	Lucasfilms, Steven Spielberg	Paramount, Brian De Palma	20th Century Fox, Brian Singer
Leading elder(s)	James Bond (Sean Connery): In a betraying body. Q: Technical toys genius, hybrid mad scientist/eunich.	John McKittrick (Dabney Coleman): Technology advisor to U.S. Defense Dept. Stephen Falken (John Wood): Coding pioneer, smart enough to listen to an intelligent teen. Gen. Beringer (Barry Corbin): Crusty technology skeptic, a "man's man."	Dr. Emmett Brown—"Doc" to young hero Marty (Christopher Lloyd): Ahead-of-his-time mad scientist, inventor of time travel.	Dr. Henry Jones Sr. (Sean Connery): Tweedy professor who still has it where it counts, always a surprise to Indiana. Once-neglectful father who learns the true meaning of family.	Jim Phelps (Jon Voight): Patriarch of the spy unit. "Max" (Vanessa Redgrave): Greedy spy for the other side, inappropriately oversexed for a "mature" woman, flirts shamelessly with the nubile Tom Cruise character.	Mutants Charles Xavier (Patrick Stewart), advocating nonviolent protest for mutant rights and Eric "Magneto" (Ian McKellen), who favors guerrilla tactics to stop persecution. Both lead young teams of assistants, but Xavier is a mentor and Magneto is a demagogue.

contd.

Table 6.2 *Continued*

	NEVER SAY NEVER AGAIN	WAR GAMES	BACK TO THE FUTURE	INDIANA JONES AND THE LAST CRUSADE	MISSION IMPOSSIBLE	X-MEN
Gender lessons	Powerful women are freaks. Good-looking women need protection.	Middle-aged men must turn over the reins to the future's virile geeks (also men). Girls are cute but insignificant to work.	Girls are cute but insignificant to work.	Good-looking women are dangerous (even if they're smart), no matter what your age. Every man is a potential victim, no matter what his age.	Middle-aged men are corruptible and must turn over the reins to the future's virile geeks. Good-looking young women might be dangerous. Oversexed aging women must be contained.	Everyone is a freak. Gender just complicates things. Humans work best in teams.
Elder's work	Save the crown, the world, and the girl. No one else can do it.	Keep the world safe from accidental nuclear war.	Implement time travel.	Find the Holy Grail, a sacred lifelong quest.	Phelps: Keep the free world safe by protecting top-secret spy list. Max: Get the spy list and make a fortune.	Xavier: Save mutants from oppression while saving the world from Magneto. Magneto: Save mutants by destroying humans.

| **Technology** | Prototypical virtual reality war game, mastered by the younger geek and learned by Bond. Artificial eye to hack into nuclear weapons command center. Laser disguised as a watch. Personal bomb hidden in a patriotic pen. Trick motorcycle. Yacht's communication command post. Lots of scuba gear and experimental vehicles for underwater and air travel. Technology equals toys. | Joshua, an ARPANET-enabled computer that learns. Technology is dangerous if not controlled by good geeks. | "Flux capacitor" housed in a Delorean, prototype time-travel vehicle. Electric guitar, unleashed: Province of the young only; the older generation just doesn't understand. | Old technologies: ancient clues and secret crypts; the Holy Grail—all used as tools for illumination. "New" technologies: Pre–WWII-era war technology (fighter planes, tanks, rifles, pistols), zeppelin—all used as tools for conflict. | PC and disks to capture spy data, classified search engine, email, high-security computer network protected inside CIA vault, spy gadgetry, such as exploding chewing gum, Dick Tracy wristwatch communicators, retinal scan systems, and spy prostheses. Disguise prostheses, used by the young geek to impersonate older characters. Jim Phelps runs his spy team remotely with elaborate tech gear. | Xavier: Telepathic-enabling, orb-like room with electronic hardware. Telepathy enables him to co-opt minds and bodies of others as prostheses. Magneto: Radiation-emitting hardware that inflicts mutancy on unsuspecting human victim. |

contd.

Table 6.2 *Continued*

	NEVER SAY NEVER AGAIN	WAR GAMES	BACK TO THE FUTURE	INDIANA JONES AND THE LAST CRUSADE	MISSION IMPOSSIBLE	X-MEN
Supporting elders?			Humorless high school principal, mistakes technological experimentation as good-for-nothing slothfulness.	Bad-guy American capitalist and Nazi officer: Both give the Jones boys a good fight but are not up to the challenge. Museum curator Marcus Brody (Denholm Elliott): Eunich-like helper/victim. Grail knight: 700-year-old guardian of holy object of faith.	70ish senator (impersonated by Cruise), provides docile persona for undercover work. 50ish spy chief Kittridge embodies cutthroat CIA spirit.	Sen. Kelly (50-something). Wants to act as Gestapo to "register" all the mutants as a means toward surveillance and behavior control

Outcome	Bond saves the girl (and the world) and vows never to return to Her Majesty's Service.	The young geek saves the world, gets the girl, and earns the respect of the middle-aged VIPs. The elder genius programmer shares the stage with him.	The young geek and the mad scientist save the young geek's world and the mad scientist's life together. Geek gets the girl.	Greedy Nazi sympathizers die violently. Brave Indiana masters ancient technologies and saves Dad. Indiana and Dad find illumination together and leave the chalice behind.	Young geek saves the U.S. spy network. Bad spies ("girl" and elder Voight) get killed.	Sen. Kelly gets turned into a mutant and dies from radiation; Magneto gets vanquished and captured; Xavier recovers from life-threatening use of telepathy machinery.
The world we live in	Virility never dies when it is used for good to triumph over evil. Young geeks, eat my dust.	Reality and game playing have merged. The young must lead the way but require guidance from a noble elder.	Anything is possible. Use caution when messing with reality. The young must lead the way but require guidance from a noble elder.	Believe in good and you will triumph over evil. The young must lead the way but require guidance from a noble elder.	Age promotes corruption. Youth triumphs.	Anything is possible. Believe in good and you will triumph over evil. Age means experience but not necessarily wisdom or corruption.

heavier and slower than when he had shot *Diamonds are Forever* twelve years earlier. The film was a pedestrian remake of 1965's *Thunderball*, but it is significant because of its ironic treatment of the aging Connery/Bond's body.

It is commonplace to hear people say or to read reviews that take the position, "Connery is the *real* Bond," as if to say that his creation and enlivenment of the role brought it an authenticity that latter Bonds have lacked. To his continued stardom, Connery has brought a persona imbued with the 007 magic, as in this notation on the Amazon Web site for the United Kingdom:

> Always unpretentious as an actor, he seemed destined to play interesting variations on the theme of sardonic sexuality. What is remarkable, however, is the increasing depth that his basic persona has acquired with age. Always good to look at, with enough irony to translate good looks into playful sexual danger, his later screen presence has developed the contours of a landscape embedded with history.[11]

Never Say Never Again banked on the proposition that the movie-going public once and forever identified with Connery as Bond, and the production winks frequently at its insider audience. For example, a new ("improved?") M, the head of the 00 spy agency, delivers early jabs in the movie after Bond fails a field exercise. "Is your edge sharp enough?" M asks. "Too many free radicals—that's your problem," indicating that Bond (Connery?) has let himself go through excess. When Q, the technology creator who gives Bond his field toys, explains that his new watch is really a laser, Bond asks how long it would last. "At least for *your* lifetime," the old man replies.

For all these yuks, Bond/Connery maintains his vigor in the story. When Fatima Blush (Barbara Carrera) crashes into him on water skis as he stands at a dockside bar, she cries, "How reckless of me! I've made you all wet!" and then stares at a big water spot on his crotch. "Yes," he replies coolly, "but my martini is still dry."

The key moment in the film is when SPECTRE agent Largo (Klaus Maria Brandauer), who appears to be in his late 30s or early 40s, challenges Bond to a virtual reality game of World Domination. The two are positioned in a casino hall at a long dining table, its Queen Anne legs protruding at either end. In between is a clear screen, equipped with a liquid display of maps and targets. Bond and Largo compete with joysticks to kill each other's armies and capture the target. The lagging player receives a current of electricity through the joystick and lets go when he can no longer withstand the pain. Bond loses the first three rounds, ending up in a heap on the floor. He comes back and wins the final challenge, defeating the younger adversary when they play for "the world." The younger, evil agent owned the technology, but Bond was up to the task of mastering it as the stakes rose. The main story has a similar conclusion when Bond defeats Largo in his bid to threaten the world with hidden, underwater nuclear warheads. In the last scene of the film, Bond/Connery (wearing only tight swim trunks) and Bond girl Kim Basinger are relaxing in a lush tropical Bahamian setting when M's emissary comes to fetch him back to Her Majesty's Service. "Never again," Bond resolutely replies. "Never?" the Bond girl asks him, followed by a kiss and then, for us, the audience, an ironic wink from 007.

How has the "son" become the "father"? First, intertextually, the younger Moore has become Bond, for the purpose of the Bond canon. This bastardization of the Bond role with Connery in the leading role underscored this advance through its acknowledgment that Connery had aged. *Never say never,* but if Connery ever does return in the role, the concluding wink seems to tell us, that role will continue to be a burden for him as he continues to age. And that is the way that the film most fertilely illustrates the theme: Through his visible aging, Bond/Connery has grown from the "son" of an ideal England to its "father," at least in the movies. He maintains a level of legitimacy and authority that few actors ever attain in their roles (his martini is still dry), but he is, finally, too old to make the stunts believable. *They don't make them like that anymore.*

War Games

This Reagan-era Cold War thriller acknowledges the dominance of the young over the forthcoming technology revolution. A young Matthew Broderick is a teen hacker. He is smart enough to break in, using a backdoor hacked password, to the computer that controls the U.S. nuclear missile silos. Both he and the computer "think" they are playing a game, but it turns out that the missiles are real, and the teenager must prevent World War III with the help and interference of three patriarchal figures: technology adviser to the military (Dabney Coleman), the crusty general who serves as the Pentagon's key player in the drama (Barry Corbin), and the coding pioneer turned disillusioned hermit who developed the computer (John Wood).

War Games paints an accidental nuclear winter as a real threat in much the same way as other dramas of its day (such as ABC Television's *The Day After*). The film's key contribution is its linking of new technology with youth. None of the three older characters either causes or can cure the crisis alone, or even through generational collaboration, because they are incapable of working together. The elder programmer, however, supports the teenager both emotionally and intellectually near the end of the drama so that young Broderick can save the world more or less on his own.

Back to the Future

Dr. Emmett Brown (Christopher Lloyd) single-mindedly works to develop time travel, even though the rest of the world thinks he's crazy. Young Marty McFly (Michael J. Fox) believes in him, though, and the two partners test

"Doc's" invention, the "flux capacitor," that Doc has housed in a Delorean. The focus of the film is on Marty's life, not Doc's. Marty's existence is threatened when he accidentally disrupts his parents' first meeting in 1954. The main problem of the film is for Marty and Doc to set things right, ensuring the future of the younger generation of McFlys.

Lloyd portrays Doc as a classic nutty professor, bedraggled, unkempt, absent minded. He is kindly toward Marty, eunich-like in his role. The sexuality that surrounds him in the film is completely outside his character, involving couplings of young McFly men and their girlfriends exclusively. Marty is appropriately respectful toward Doc's intellect, and the scientist works patiently to impart in Marty an appreciation for the process of his craft. Marty is Doc's understudy, and, because no one else will listen to the "mad scientist," we leave the film with the sense that the future will someday be in Marty's hands. Marty's own father is oblivious to the boy's developmental needs, by contrast, so Doc is truly his father figure if not his genetic parent.

Indiana Jones and the Last Crusade

The third of Steven Spielberg's Indiana Jones films paired the younger archaeologist Dr. Jones (Harrison Ford) with his father, Dr. Henry Jones, played by Sean Connery. "Indy's" adventure this time is to track down the Holy Grail of Christ and in the process find his kidnapped father. The two men are a visual contrast: the ruggedly handsome Indiana, clad in explorer khakis with all the accoutrements, including a whip, and Dr. Jones Sr., tweedy professor, umbrella in tow. Indy is "today's man" of 1938, the film's setting—driven by testosterone to the Earth's edges. He boyishly climbs out his office window to escape a line of students whose work he has not graded, having thrown his attention to yet another archaeological adventure. His father is yesterday's man, all Cambridge, a man of letters.

Dr. Jones Sr. does have a few technical tricks up his sleeve, however. He uses the wartime technology of 1938 as ably as his son to share in their fight for escape, and he brilliantly employs the low-tech bumbershoot to scare a flock of seagulls off the beach, wrecking the propellers of an airplane that is after the two Joneses. (This scene ironically recalls Connery's role as Bond, whose technology was daring for its time but, as we watch him again and again on video, contrasted with 007's newer incarnations, is sadly dated.) Chiefly, though, he has tracked down the grail through a series of opaque clues he has collected through painstaking research, culminating in the revelation of three life-threatening obstacles that finally stand between Indiana and the hidden grail. Crucially, it is the younger Indiana, not his father, who must fetch the grail from its hiding place. The elder Jones has been shot by the bad guys and must drink from the grail in order to save his life. He has provided the intellectual labor to help Indiana through obstacles, but the younger man must undertake the strenuous journey through the treacherous labyrinth while the elder one lies on his back at death's door.

By the film's end, the elder Jones has earned his son's respect as a parent after declaring his love for the son. Indy has earned Dad's respect as an archaeologist, even though his rough-and-tumble quest methods seemed impatient and imprudent to the father. They leave the scene, having fulfilled the father's long search for the grail (which they had to leave behind anyway). The father's appetite satiated, it is understood that the son will continue to venture forth.

Mission Impossible

The action thriller has two leading elders—Jon Voight as Jim Phelps (good guy turned bad guy) and Vanessa Redgrave as Max (bad "guy"). Tom Cruise is the young star, the No. 2 player in Phelps's spy ring, Ethan Hunt. Phelps dupes Ethan in a double sting that implicates him, but Ethan brilliantly dupes Phelps back, saving the free world in the process. He works in league with world-class freelance information dealer Max, Vanessa Redgrave, who pathetically flirts with Ethan. Ethan and his young associate deliver Max to U.S. spy agents in the end, taking her off the streets as well.

During the movie's climax, Ethan humiliates Phelps by revealing to him how he unraveled the elder agent's elaborate ruse. Ethan doesn't just capture Phelps, he kills him, after a lengthy and violent struggle that is packed with special effects.

Phelps has been Ethan's mentor, directing him in the deployment of the most sophisticated spy technology available, from high-security search engines and latex disguises to spy eyeglasses and fingerprint security entries. At first, Ethan uses these technologies in Phelps's employ, unwittingly advancing the crooked agent's plot. Through the first portion of the film, we see Phelps running the controls, seated at a remote computer, monitoring the operation, dispatching orders to the ring of agents in the field, with Ethan as his field deputy. Later, though, Ethan re-employs the technologies to defeat Spymaster Phelps, improving on the old man's mastery.

X-Men

Two aging actors play the lead roles in this science fiction action film from the summer of 2000, but their age is overshadowed by their status as mutant hero and mutant anti-hero of the story. We meet them both at a Senate hearing in which a McCarthy-like politician is campaigning for the outing of a growing number of mutants—people whose biological makeup has been transformed at puberty by the radical addition of certain extraordinary powers, such as the ability to walk through matter, transform molecules into fire or ice, or move objects. Dr. Charles Xavier (Patrick Stewart with ten years' exposure as Jean-Luc Picard of *Star Trek: The Next Generation* series and movies) is the voice for peaceful progressiveness in behalf of the mutants of the "not too distant future." His "School for Gifted Children" takes in human mutant runaways, training them to discipline their extraordinary talents for good. Xavier and a team of "X-Men," a few of his star alumni, peacefully oppose a U.S. Senate move to require mutants to register with the government. Ian McKellan, also cast as Gandalf in *The Lord of the Rings*, is Xavier's nemesis, Magneto, who directs Xavier to "get out of the way" as he brings a menacing, violent opposition to the mutant registration effort. Both characters have extreme powers, which they control masterfully: Xavier is a telepath who can read others' thoughts and control them; Magneto can control metal, as we learn in the opening scene, when he is taken from his parents as a young teen in Poland and brutalized in a Nazi death camp. Magneto's orientation toward his mutant status and the politics of discipline that engulfs it is steeped in his Nazi Holocaust experience: He fears that required mutant registration is the gateway to unfettered marginalization and retaliation by "normal" humans, whom he sees as the real danger.

Among the ensemble of supporting characters is a teenage girl, Rogue, whose "gift" is to drain the life force from any human being she touches, effectively ruling out human contact for her forever. Rogue easily allows us to see the vulnerability of the adolescent, faced with the aloneness of existence in a harsh social world that has been compounded by bigotry and conflicted emotion. Nemesis Magneto is similarly torn. Unlike villains of most action films, he is not a truly "bad guy," only a victim who favors harsh methods to protect his and other mutants' lives. Because we first meet Magneto when he is roughly Rogue's age, and because he chooses her to unwittingly help him accomplish his goal of vanquishing the controlling human "normals," it is simple to associate Magneto's struggle with the difficulties of adolescence. The movie, like the comic book series that inspired it, clearly means to address teenagers as its primary audience, making it especially interesting that the authority figures we must choose between are both gray-haired men. Especially interesting is Xavier's status as disabled. His wheelchair use is treated unremarkably in the film, and instead the story underscores his humanity and "cerebral" capacity.

Xavier's technology, apart from his wheelchair and raw telepathic ability, is a special, spherical room that we learn he constructed with the help of Magneto when the two were friends. This cavernous space is equipped with a suspended catwalk and metal headset, which extraordinarily extends Xavier's ability to "see" telepathically. He is a disciplined and powerful user of the technology. His young and attractive female student, Dr. Jean Grey, is not allowed to approach the grand technology yet, because she is "undisciplined." Only at a critical moment in the film, when Xavier is rendered comatose, does Jean slip into the room to assume the role of hero through her risky engagement with the medium, and she is knocked off her feet by the headphones' signal.

Magneto's technology is appropriately menacing. By transferring his own body energy, he forces a giant windlass-like device to spin around a human's body, capturing the powers of his victim through radiation, the latter of which is lethal. Magneto's first captive is Sen. Kelly, the McCarthy figure.

Late in the film, after Xavier winds the first round of what will be a multiproduction battle with his arch-rival, he and Magneto are shown playing chess in a cell furnished all in glass and Lucite. Obviously, the authorities knew that it was too risky to place Magneto in a metal environment or to give him metal tools or toys to play with. Moving lesser pieces across the board toward each other's king, Xavier and Magneto allegorically play out the drama they have just conducted, each having worked against the other with a band of young mutants employing their specialized technical body knowledge for pure good or for evil-good.

While its storyline is unambitious, *X-Men* deftly takes a stand for nonviolent protest against the horrors of prejudice while unsentimentally acknowledging that personal experiences leave room for more severe oppositional reactions. With Xavier positioned as the film's hero (more or less, as he is positioned amid an ensemble cast), it is heartening to see a disabled elder, at least in his 60s, snowy white chest hair and all, spurring the drama along. With Magneto as the ambiguous anti-hero, we see a complex, well-meaning elder in the rival position. Both these characters, created by actors trained in Shakespearean performance, are far more eloquent than any other in the production and lend the notion that one grows to old age rather than being reduced to it. That neither is "retired"—we see them entirely at work—underscores this conclusion.

The ways that the hero and anti-hero of the film deal with their respective teams of understudies differs sharply. Xavier is a patriarch with an unselfish eye on the future. He devotes his existence to training his replacements, the young adults who form the "X-Men" squad: Dr. Jean Grey, Storm, Cyclops, and, it is implied, Wolverine, along with raising a crop of "gifted children" at his boarding school. He takes pride in their individual achievements and particularly concentrates on preparing Jean to take his place as chief telepath. That her talents extend to telekinetics—an arena where he has no gift of his own—implies that his "child" will grow to exceed him in his capacity as leader, something most parents freely admit that they wish for.

Xavier's counterpart, Magneto, treats his team of young assistants as mere henchmen. We know nothing of their personal stories, and they exercise almost no agency in their dealings with him. They are flat characters whom we are not meant to see as possible successors to Magneto's throne. They are expendable and will not be missed if they fail to appear in sequels.

What distinguishes *X-Men* in the "Child Becomes the Father" category are that, first, the "children" are not all male, and, second, unlike most action films, the exposition emphasizes the elder characters rather than merely the younger ones. In the Indiana Jones movie, for example, the perspective is clearly from Indy's point of view. We follow that character through almost the entire film and we see Dr. Jones Sr. largely from Indy's point of view. The case is similar for *War Games, Back to the Future,* and *Mission: Impossible.*

When we see the younger characters in *X-Men*, we can be reasonably certain that special effects and action will be part of the program. For example, at the climactic moment near the end of the movie, a long battle scene occurs at the Statue of Liberty. Magneto is in the foreground along with his potent young helpers, but Xavier is back home at the mansion, lying comatose on a table. The team of four young X-Men are there, along with the teenaged Rogue, to defend Xavier's cause and vanquish the evildoer. Then, in the following scenes, Xavier comes to the forefront again, first to express his approval at the young team's accomplishments and growing good judgment, then to face off against his nemesis in the chess game in Magneto's clear cell. We are assured through this resolution—which is only partial, as it implies a rematch—that the two arch-enemies are still very much in charge of their conflict, that it remains personal, and that they will continue to direct young squads to deploy their strategies. Xavier gives up ground to the X-Men in the story but does not step aside.

Revisionist History

Often an elder character's job in a film is to drive home the point that reality isn't as we'd thought, or things aren't what they used to be. These revisionist tales of modern reality sometimes come with a light touch, sometimes not. It is generally the older character's job to illustrate that the old ways of work have given way to new ways, new technologies, and a new generational paradigm. The films I have chosen to illustrate this theme are a screwball comedy from 1957 and five later twentieth-century movies, all from major studios (see Table 6.3).

The Desk Set

An early representation of the aging/work intersection with new technologies is the 1957 screwball comedy *The Desk Set,* the final pairing of Katharine Hepburn and Spencer Tracy in a romantic comedy. The film depicts the installation of computers into the American office in the name of twentieth-century efficiency. Hepburn plays a 40ish mother hen to a television network reference library, staffed by a small pink-collar crew who answer phone calls from harried, unseen creative workers (males in need of helpmeets) in need of quick facts. Hepburn runs the operation with competence, even if it sometimes takes weeks to find data. Then comes Miss Emmy.

Tracy's character barges in as an efficiency engineer whose job it is to install Emilac, a newfangled, industrial-size artificial brain, with its dazzling array of blinking lights, shooting cards, and amazingly fast memory. Also entering is a young, humorless woman of science, who lets Hepburn know straight away that the office must be wiped clean of her old-fashioned ideas. It appears that Emilac will render the human crew redundant.

It is significant that Tracy is already white-haired in the film and appears perfectly "natural" as the modernizing agent. He acts logically, ultimately imposing a lesson on the sentimental women: Technology and humanity must work together in the name of efficiency. The women are then free to empower themselves through the computer's labor so they can adopt new skills and improve their jobs. Near the film's end, the computer goes haywire and, of course, Hepburn is free to respond grittily: "Come on, girls. Let's show 'em what *people* can do!" The team triumphs over adversity and its wisdom is rewarded. Technology is meant to complement humans, not to render them inferior.

The film's representation of age is problematic, though. Viewers are left to presume characters' age, and the mature Hepburn portrays a coquettish character who swoons and giggles, all through a soft-focus lens. She is the voice of wisdom without the trappings of a sagging body. Middle-aged assistant Peg (Joan Blondell) is Ethel to Hepburn's Lucy. Past her sexual prime, Peg is a "career girl."

The film inscribes its times with a couplet of patriarchal truths: Men know best about big things, as Tracy's character demonstrates when he reveals the working conditions that Emilac will bring. Women, though, know best about small things, and that informal power really runs the world. This feminine knowledge is constructed on the terrain of intimacy, as when Hepburn deduces that Tracy is single because of his unmatched socks, or when she supplies an emergency "tool" for repairing the computer: her hairpin. *The Desk Set* leaves us with a conclusion that only partially holds up more than forty years later: When aging workers are in charge, we still expect them to be men. But it is unlikely that a *Desk Set* for today would showcase a white-haired IT engineer in the role of change agent. More likely, the character would be under 30. Following Hollywood rules, that would render all but the youngest starlets too old for the female role.

The Conversation

Harry Caul, renowned in the "surveillance and security technology" field, celebrates his 44th birthday in the early moments of this 1974 Francis Ford

Table 6.3 Revisionist History Films

	THE DESK SET	THE CONVERSATION	DRIVING MISS DAISY	STAR TREK: FIRST CONTACT	ENEMY OF THE STATE	SPACE COWBOYS
Release date	1957	1974	1989	1996	1997	2000
Studio and director	20th Century Fox, Walter Lang	Paramount, Francis Ford Coppola	Warner Bros., Bruce Beresford	Paramount, Jonathan Frakes	Touchstone, Tony Scott	Warner Bros., Clint Eastwood
Leading elder(s)	Richard Sumner (Spencer Tracy)	Middle-aged Harry Caul (Gene Hackman), socially inept geek caught in sting by slick young tech experts and adulterers.	Miss Daisy (Jessica Tandy), an elderly Atlanta Jewish matron; Hoak (Morgan Freeman), African American of few words who becomes her chauffeur in middle-age and grows into his own elder years as her servant and friend.	Jean-Luc Picard (Patrick Stewart), commander of the Enterprise. Picard becomes emotionally entangled with an Earth woman from the 21st century; a scientist working with a spaceship pioneer.	Reynolds (Jon Voight), murderous, power-mad National Security Agency middle manager; Brill (Gene Hackman), independent anticorruption communication analyst.	Old flyboys (Clint Eastwood, Tommy Lee Jones, Donald Sutherland, and James Garner) reunite to fly a space shuttle mission to correct a programming error in a Russian satellite. Originally denied astronaut status, these elders are in demand because only they have the technical knowledge necessary.

Gender lessons	Women have all the (punch) cards, but it takes a man to run the computer.	Technological expertise is no match for adulterous greed.	Women are especially dependent when they are old.	Women and men work together, but men are in charge.	Women who play with technology fire risk getting killed. Men are most vulnerable about matters of fidelity.	Women make fine scientist/doctors and wives, but real men fly space ships.
Elder's work	Efficiency expert, introducing computer into pink-collar research library.	Use surveillance technology to catch bad guys.	Daisy: Remain vital, despite her disobedient body; Hoak: Drive Miss Daisy.	Pitted against the Borg Queen to save the universe from assimilation into the hive.	Voight: Use technology for evil and personal gain, befitting the National Security Agency's goal of unbridled technology surveillance power. Hackman: Beat the bad guys at their own game, by using technology to catch them cheating.	Right a renegade satellite.

contd.

Table 6.3 *Continued*

	THE DESK SET	THE CONVERSATION	DRIVING MISS DAISY	STAR TREK: FIRST CONTACT	ENEMY OF THE STATE	SPACE COWBOYS
Technology	Programming knowledge (no match for Hepburn's bobby pin and razor-sharp mind). Technology is good when it's understood and in the hands of a benevolent force.	Sophisticated (for the time) surveillance technology: aggressive bugging and spy camera devices. Technology is an index of humanity's decline. Power corrupts through technology.	The car: As Daisy's prosthesis, it is first the symbol of her independence, then her dependence. As Hoak's prosthesis, it is his leverage for controlling his master, Daisy.	"Old" technology: First ship to fly at warp speed. "New" technology: Enterprise, phasers, holographic communication, droids, cyborg hardware.	Sophisticated surveillance gear turns life into gritty videogame panopticon. Power corrupts through technology.	The Space Shuttle, a satellite, and (from the past) sound-barrier-breaking aircraft.
Supporting elders?				Zephron Cochran: Weak of character, he becomes convinced that his mission (flying the space ship) is more important to humanity than his own fears are to himself.	Congressman (Jason Robards) opposing loosened telecommunications policy for security agencies. Murdered by Voight's geek minions in first scene.	Mission control pal (William Devane); Old Air Force nemesis, now with a command post at NASA.

Outcome	Technology (Tracy) and humanity (Hepburn) marry after a struggle that ends in compromise.	Caul's life is ruined. He can never escape surveillance. He tears up his apartment searching in vein for bugs.	Daisy decays until she must retreat into a nursing home. She tells Hoak he is her best friend.	Cochran's rocket flies, the Borg are contained, and the future is saved. Picard triumphantly restores his reputation.	Young lawyer Robert Dean (Will Smith) and Brill team up to stop Reynolds and company through high-tech sting.	Young astronauts almost wreck the mission, but Space Cowboys save Earth when they wrangle weapons of mass destruction away from the satellite and shoot them to the moon, strapped to a terminally ill Tommy Lee Jones.
The world we live in	Magic (technology) is good, as long as you listen to a good white man. Women are emotional and foolish about technology.	If the bad guys want to hurt you, you cannot stop them.	Axes of power (gender, race, age) make the world a confusing place. Love is really all we have.	Peace is worth fighting for. Progress is peace.	Tension between need for government security and privacy for citizens leaves unsettling dilemma about technology policy. If the bad guys want to hurt you, you probably cannot stop them.	Experience counts. Rugged individualists can work in small groups to use technology more effectively than green horns can. Viva John Glenn.

Coppola film. Gene Hackman, in a conservative suit and filmy nylon raincoat, plays the secretive Caul, running a team of surveillance operatives who are taping an adulterous conversation by a young couple walking in circles around a noisy park. The woman, married to a rich executive many years her senior, and her lover appear to be talking of their fears that the executive might kill them both. Instead, as Harry learns through further, unofficial surveillance, the murder plan is the other way around. "The conversation" was a setup, designed successfully to lure the husband to a motel where he thought he would catch the young lovers. Instead, they catch him. The wife, her lover, and an accomplice, the executive's assistant (Harrison Ford again, this time quite young), kill the rich man and take over his company. The young trio deliver a warning to Harry: They know he knows about them, and they are "listening" to him by turning the surveillance on his own life. At the end of the film, Harry sits, playing the saxophone, in his apartment, which he has torn apart from ceiling to floorboards searching in vain for bugging devices. His tools have been turned on him, and he is a defeated, impotent man.

Harry Caul, apart from his technical involvement in the murder, is also a character study unto himself: He is pathologically private and crosses increasingly into the realm of paranoia. He is so careful, in fact, that he loses his girlfriend (a 20-something Teri Garr) rather than let her know about his private life. This protectiveness toward personal identity is a wonderful irony, considering Caul's wiretapping business. He is busy listening to other lives being led but it mortifies him to think that someone might get too close to him. Meanwhile, the voyeuristic Caul rationalizes to struggle with his own work's effects on others' lives and losses.

The film is a dark comment on technology, a cautionary tale about the postmodern world that Coppola observed three decades ago: Privacy as we once understood it is over. If people want to get you, they can. You are helpless. Secondarily, we see the technology grasped confidently and menacingly in the hands of the young. The elder executive is killed, and Harry, the so-called expert, sees his own life in ashes. Harry is a master of analog audiovisual taping—he can bug anybody—but he is a dupe. His work, which brought him great accolades, is meaningless and even, for him (a devout Catholic), the source of unconscionable sin, because he has unwittingly brought harm to others through it.

Driving Miss Daisy

Daisy Werthen (Jessica Tandy) is elderly and Hoke (Morgan Freeman) is younger but gray-haired when the two meet. Daisy's son has hired Hoke to serve as her chauffeur after she drives her car onto a retaining wall. She resents this new servant as the symbolic loss of her independence. She acquiesces after a struggle and works to maintain her dignity and her place in her community as a Jewish mill matriarch in mid-century Atlanta. She needs the technology—the car—that only he can deliver to her, because the insurance company will no longer allow her to drive, and she must get about. Over the next twenty-five years, she and Hoke become close. Near the end of the movie, he reports to work in the morning as usual and finds her in a confused state. After he calms her, she holds his hand and looks into his eyes, saying, "You're my best friend, Hoke." In the film's last scene, Daisy's son has brought Hoke to the nursing home to visit her on Thanksgiving in her final days. She is a very old woman, and he is past driving anyone, including himself. Hoke observes that Daisy has not eaten her pumpkin pie. She struggles with the fork before he takes it over and begins to feed her.

The film demonstrates the complexities encountered at the intersections of age, race, ethnicity, and social class, highlighted by the setting in the Deep South. Daisy cannot drive herself, because she is too old. Her servant (it is unstated) can only be a man and, at the time, only a black man. It is the not-quite-New South, a time when Hoke can drive Daisy all he wants to, but he is not allowed to use the restroom at the service station when they stop for gas. They are treated as an oddity. Two state troopers who hassle them about their business in South Alabama remark wittily as Hoke and Daisy pull away in the black Cadillac: "An old nigger and an old Jew woman taking off down the road together. That is a sorry sight." Daisy and Hoke struggle with their professional and personal relationship because of their social class and black/white difference. She has him drive her to a Martin Luther King Jr. rally for civil rights. She proudly attends with all the other good, white liberals and a smattering of African Americans, who, seated during the talk, are ringed by red-jacketed black waiters standing in the background. She wears a fur stole to the dinner and sits alone, an empty seat next to her that had been intended for her son, a successful young Jewish businessman who had decided that his attendance might offend white Christian business associates. Hoke sits in the car, listening to the King speech on the radio.

The night is part of Daisy's own education through Hoke. Another is when he tells her, "I am not some back of the neck you look at when you have to go somewhere. I am a man." Hoke may operate the technology, but he is not the tool. However, in mid-century Atlanta, he also lacks full agency, and he knows it. "Yas'm," he tells Daisy when she asks for something, even if he jokes freely about her to her elderly black maid.

For Daisy, the car and the man begin as the same character. Eventually, the cars are all gone, and she realizes that the man is someone much more to her than the transportation had ever been.

Star Trek: First Contact

Two elders are at the forefront of this film. The Enterprise's Captain Jean-Luc Picard (Patrick Stewart) and Zefram Cochrane (James Cromwell) have no interaction, but the story turns on their interrelated struggles with technology. Picard must rescue the *Enterprise* from the Borg, the warring cyborg communal species that has captured the ship in its attempt to colonize Earth and disrupt the struggle for universal peace. Picard's nemesis is the Borg Queen, a fusion of humanity and technology who leads the venture. Picard ultimately unleashes a plasma substance that destroys the organic component of the Borg enemies on the ship. In doing so, he also must take away the organic skin and nerves that the Borg Queen has grafted on to Data, the ship's android and intelligence analyst. This is a bittersweet moment, because vanquishing the Borg saves the *Enterprise* and the Earth, but Picard knows that it costs Data a significant step in fulfilling his desire to be human. As with all of Picard's decisions, however, he intelligently employs technology for the greater good of humanity, never in the service of his individual wishes. Picard is the embodiment of the Star Trek ideal: peace through enlightenment and understanding.

Cochrane is a historical figure from the twenty-first century, which the *Enterprise* officers encounter through time travel in pursuit of the Borg. In his late 50s or perhaps 60, Cochrane is out drinking on the eve of his maiden flight of a space ship that will, according to the history books, be the first to travel at warp speed and thus attract a visit from an intelligent species who observe it. The timing of the mission is critical, because the extraterrestrials' ship will pass through the region only on the following morning.

Meanwhile, Cochrane's plans are derailed by the Borg, who stage an assault on his community and damage the ship. Cochrane retreats into the bottle. *Enterprise* staffers are saddened to learn that not only is he an alcoholic but also that he is not especially interested in space travel, only in the money he can get from it. They rehabilitate him and he finally hears his noble calling in the name of humanity. The "primitive" space craft flies successfully, with some "mental" help from Cochrane's rock-n-roll disk, and Earth gets a goodwill visit from a crew of Vulcan space travelers. An era of peace is ushered in. Once again, communal rather than individual values win out.

Enemy of the State

Jon Voight and Gene Hackman play two opposing roles for older men working with technology in this film, which stars the affable, young Will Smith. Voight, as Reynolds, is a power-mad murderer set loose in the National Security Agency (NSA). He uses surveillance technology ruthlessly, supervising a team of Gen X geeks to ensnare Robert Dean (Smith) in an evil plot. Dean finds Brill (Hackman), a disillusioned NSA dropout who lives hermit-like, a bit like Harry Caul, in an abandoned warehouse where he eavesdrops for money on evil communications throughout the world. Brill coaches Dean in a crash course on surveillance so that he can free himself of the bugging devices that Reynolds's team has unleashed on him. A climactic fight underscores the potent Dean, not the 60-something Brill, in the lead role of hero, and the bad guys are vanquished.

Brill, like Harry Caul, is a surveillance and digital file-capturing god, whom the younger player Dean marvels at for his expertise but cannot fully embrace because of his solitary, secret personal life. Dean is surprised near the end of the movie when Brill reveals that, out of duty to a late friend, he had been "looking after" his and Dean's mutual young female acquaintance, played by Lisa Bonet, who is murdered by Voight's team of sociopathic geeks. Dean discovers that Brill is deeper than he had thought.

The younger Dean clearly counters the elder Brill as a virile action hero, but the film acknowledges the value of experience and serious study as the province of the older worker. It calls into question the popular notion that older techies are obsolete and cannot expect to profit from a free enterprise system that rewards merit and guile. Reynolds teaches us only that power corrupts and that power tends to come with age. When his young team of geeks follows him into explosive defeat, the film preaches a critique of youth's potential blind acquiescence of power, no matter how it might allow the geek to flex his own technically creative muscle.

Enemy of the State is about a corrupt world in which lawmen and cowboys work together to reinstall legitimate authority. That young and old find methods of collaboration in the film is a tantalizing proposition, especially since the film's real hero is Smith, who is not white, and the "buddy" of this odd buddy picture is an old white guy. Significantly, the women in the film are vulnerable and do not play at the highest level of the action, and they are quite young. I am left wondering how the film might have been different if *Mission: Impossible*'s Redgrave had been cast as Brill in this particular movie. It seems still outside the realm of Hollywood possibility that an older woman can be depicted as truly potent in the technology arena.[12]

Space Cowboys

In 1958, Frank (Clint Eastwood), Hawk (Tommy Lee Jones), Tank (James Garner), and Jerry (Donald Sutherland) were the best pilot team the Air Force had, and they fully expected to be the first men in space. But their commanding officer, Bob Gerson (James Cromwell), bypassed Team Daedalus's four would-be astronauts and established the fledgling space program without them. In 2000, a disabled Russian satellite is in need of repair and Gerson is forced to call Team Daedalus out of retirement and send them up on the Space Shuttle to repair a broken guidance system, which had been designed by Frank for Skylab. "It was ahead of its time, but it's obsolete now," a young female NASA staffer tells Frank, and that seems to be NASA's view of the four senior citizens as well. The "rookies" prove the skeptics wrong, passing physical requirements and learning to operate the advanced technical systems associated with the Space Shuttle Daedalus. Hawk, however, is diagnosed with terminal cancer. When the young astronauts who are placed on the shuttle with the team wind up bungling the mission, it is fortunate that Hawk is there to sacrifice what is left of his life. He volunteers to release himself into deep space with a load of nuclear warheads that Hawk discovers on the satellite. Hawk dies a happy man, having saved his colleagues and the Earth and finally getting to step onto the Moon, where he runs out of air.

The hoopla in the film surrounding the "senior" astronauts echoed John Glenn's flight on the space shuttle in 1999 (designated by the United Nations as the International Year of the Older Person). Team Daedalus is celebrated by Jay Leno and *USA Today* as "the ripe stuff." Along with their technical expertise, the squad has its members' sexual prowess celebrated, too. Men will be boys, no matter what their ages, and anything worth doing is worth making into fun, the film continually reminds us.

It is especially significant that the two young, cocky astronauts cause the near-catastrophe that leads to the film's climax and that only Team Daedalus, with its collective technical experience, judgment, and fire in the belly can fix it. After losing Hawk, Frank limps back to Earth in the disabled shuttle. He and his two remaining buddies help the unconscious young astronauts parachute to safety in the ocean before they set the "bird" down on the runway at Cape Canaveral in a daring sans-instruments landing for which Frank had prepared by observing the reckless Hawk in training.

The film was directed by Eastwood, and the message was unambiguous, translatable for Hollywood and much of the rest of the working world: Celebrate the seasoned professional. The young folks need to earn their stripes. Don't send a boy to do a man's job. On a deeper level, the elder characters' anxieties about how they appear to the younger working world—and to each other—is Eastwood's acknowledgment that old age is scary. In a cafeteria scene, ponytailed Jerry mortifies his partners when he adjusts his dentures publicly, in front of the young astronauts at nearby tables. When the younger guys send over a platter of Ensure as an insult, Jerry brags that he drinks it regularly because it enhances his sexual performance. Eastwood's directorial perspective is decidedly masculine, and there is a poignancy with which he delivers Team Daedalus's internal banter. The characters compete to see who can run faster, read more tiny letters at the bottom of the eye chart, and get the bar waitress's attention. The consequences are sometimes unintentional, as when Hawk asks a waitress, "Which one of us would you take home?" She answers: "You mean to THE home? The retirement home?" Eastwood lets his characters risk the embarrassment of being dismissed as old fogeys but he imbues them with a competence that allows them to emerge victorious.

Conclusion

The three categories I have presented allow for overlap. For example, *X-Men* could easily be explained as a "revisionist history" film, as it shares more with *Star Trek: First Contact* than just Patrick Stewart. Both films and the original texts from which they spring interrogate ideas about prejudice, human limitations, and the possibilities that technology might play in liberating us from both. Likewise, I have argued that *Star Wars* is a fairy tale, but it also has overtones of "Child Becomes the Father" and "Revisionist History." Identifying the three themes, however, does allow me to make the point that film organizes the ways in which it treats elder characters, even if that organization is subject to revision and overdetermination.

Many other interesting cinematic treatments of elder characters have appeared, particularly in the last several years. Some of the most interesting of these have been *The Joy Luck Club* (daughters learning from mothers about agency), *Crouching Tiger: Hidden Dragon* (young warriors learning about agency from elders), *In the Line of Fire* (aging secret service agent admits his decay but toughs out his assignment anyway), and, especially, *Contact*. The latter movie, a Robert Zemeckis film starring Jodie Foster in the provocative role as a visionary and determined young female astronomer-turned-space-traveler, features two informative male elders. David, played by Tom Skerritt, is Ellie's (Foster's) nemesis, a gray-haired Baby Boomer who condescends to her about her research, then, when she establishes its legitimacy by proving that extra-terrestrials have sent communication to earth, uses his white-male power to grandstand for the media and steal her right to win the reward of traveling to meet the aliens. Ellie tolerates Dave's bad behavior with grace, then heroically tries (and fails) to save his life when she observes a terrorist about to kill him before he takes off for space. Young Ellie literally and metaphorically "survives" this terrible moment in order to get her just reward of traveling to meet the aliens herself.

The second elder character is much older. John Hurt plays Mr. Hadden, iconoclastic billionaire, who handpicks Ellie from afar to support her dream, which he savors vicariously. He bankrolls her adventure and supplies her with desperately needed information and tools at the critical moments, just when it appears she has bottomed out. Near the end of the film, Hadden dies. He is demonized as a hoaxter by the film's evil politician figure in an attempt to discredit Ellie, who has returned to Earth without apparent physical evidence of having encountered the aliens. But it is too late: Hadden's support has allowed Ellie to capture the public imagination, and she becomes a hero.

Skerritt and Hurt's characters show two polar opposites of elder figures in so many films I have seen: Skerritt's character represents the holding back of youth, the corruption of age, the threat of evil that we see over and over again in the evil stepmother, Darth Vader, the corrupt agents played by Jon Voight, and the skeptical general in *War Games*. In these characters, we see age as a decline, a reduced capacity, a repression of possibility.

In Mr. Hadden, however, we see a common elder figure as well: the enabler of the future. Hadden abets Ellie from afar (in a *Charlie's Angels* sort of way). Because he is wise and powerful, he can impart the future to the young visionary, removing obstacles in her path. But make no mistake: The drama is about the young, so Ellie must demonstrate her bravery and grit to legitimate herself to the audience and to Hadden.

This chapter has suggested some of the messy ways that we, as a culture, are trying to work out the balance between persistent and dynamic models of old age in a time when everyday life is subject to a parade of fresh experiences. Hollywood film has been one of the most influential storytellers of our time. It is helpful to realize that older characters frequently are invoked not because the filmmaker may wish to tell a story that involves an older person but because the older characters easily function as a sort of shorthand for facilitating the story that is being told. In many of the films discussed in this chapter, I have described tales in which we meet young characters who are in need of intellectual and emotional education, and older characters are often the vehicle for this necessary adult development. Consider Merlin's influence on Wart/Arthur, Obi-Wan's impact on Luke, Doc's work with Marty, and even Phelps's training of Ethan. We meet the older characters in such circumstances because they lead us to understand the challenges for the younger ones and to appreciate the superiority of the young heroes.

On the other hand, old age is exactly the point of some of the stories, as in the cases of *Space Cowboys* and *Driving Miss Daisy*. In today's film environment, when age *is* the point, it has to do with the identity exploration being done by Baby Boomers, whose discomfort with slouching toward decrepitude must be replaced with more tolerable narratives, such as the quest for new relevance. When someone in the country who grew up with Elvis or the Beatles is turning 50 or 60 every few seconds, that creates a strong undercurrent to stretch and challenge foundational myths.

I expect many more *About Schmidt*s to emerge in the next few years, not only because of the market (or need) to understand and digest the late-life changes for the Baby Boom, but because of the capacity of big stars such as Nicholson, Robert DeNiro (as in *Meet the Parents*), and Eastwood to both model old age on screen with finesse and retain their access to Big Hollywood. As a generation of female directors, such as Penny Marshall, ages, it will be interesting to see what new narratives about age emerge from the influential studios.

With the aging of the Baby Boomers, I am hopeful we will see more films like *X-Men, Driving Miss Daisy,* and *Space Cowboys*. These are stories in which strong elders thrive without nubile shills to draw the crowd. When a Meryl Streep in her 60s can draw the kind of audience—and money— that a Tom Cruise in his 30s can, we will have something more to talk about.

CHAPTER 7

Who You Callin' Dude?: Magazine Advertisers Discover Older Workers

Magazines adopted a purposeful approach to audience segmentation long before the broadcast networks figured out that there were desirable and undesirable demographics.[1] Still, the advertisers that pay for magazine advertising have been slow to catch on to ways of imagining the older consumer that go beyond assistive goods and services (from lift chairs to no-questions-asked life insurance policies). The 1990s saw a dearth of older computer users in images of technology-focused magazines. Magazines typically were geared toward business geeks and pictured young (usually white) males at work, or they were geared toward "families" and pictured young men and women and their young (usually white) children. Toward the end of the decade, women and people of color tended to be more evenly represented in business-related technology advertising, but they were almost always young.

I was especially struck by how the faces of these young technology worker models contrasted with the head shots and other images of the editorial subjects of the magazines—often the stereotypical white-haired CEO figure amid the trappings of executive success. These older men swam in the digital waters and influenced the tides, but they were unaccounted for by the advertising whose photographs idealized the perfect technology user at work. That person was almost always an apparent Generation Xer.

This business-to-business advertising imagery fit well into a larger universe of technology-related advertising. Consumer-focused ads, common in both television and magazines, tended to ignore the older person as a potential adopter. As the "senior surfer" stereotype became entrenched, in around 1999, ads targeted older tech consumers in specific outlets, such as *Modern Maturity*, AARP's magazine, but the image of the older computer user was one of using the computer for leisure: Email your grandkids, book a cruise on line, do your genealogy. Older workers were nowhere to be found.

In the last few years, however, that has begun to change; although older workers are by no means evenly represented in technology-related advertising, their appearance has begun to be sufficiently regular so as not to be shocking any longer. This trend obviously has been fueled by the swell of late-career Boomers in the workforce. This chapter examines images of older workers relating to technology in mainstream U.S. magazines in fall 2001. (See Table 7.1 for a list of the magazines examined and the type of ads that appeared.)

The revved-up editorial play and the advertising that made it possible were launched amid a more general push to recognize the 600-pound gorilla of aging Americans in the room of media audience-hood. For example, the *New York Times Magazine* on March 9, 1997, featured a collection of aging stories and reflections by and about famous Americans over 60— from Eartha Kitt to Ed Koch. The cover photo, which appeared on a white background, showed thirteen cultural luminaries dressed in black, recessed behind the bold-lettered headline: "Funny, We Don't Feel Old." The teaser on the bottom of the page read: "America discovers a new stage of life—after middle age. The Age Boom. A Special Issue."

Then, in July 2002, the unthinkable occurred.

Time magazine devoted two consecutive covers to gray-haired women. On July 22, the cover photo of Susan Pierres, 60, depicted a fashionably dressed, blue-eyed, silver-haired woman, shot before an ocean backdrop, who had been taking estrogen replacement therapy for a decade. The headline was in all caps: "THE TRUTH ABOUT HORMONES," followed by a subhead: "Hormone-replacement therapy is riskier than advertised. What's a woman to do?" About Pierres, the photo caption added: "She is angry and confused but not yet ready to stop taking them." In addition to the cover story, inside the magazine, the Notebook: Milestones section informs readers that seven years ago, scientists had announced that estrogen replacement therapy might be linked to increased cases of breast cancer.

The next week, *Time* aimed for a younger audience but pictured a "mature senior" on its cover. "Anna" (identified by her sewn-on label) was a cartoon-drawn figure on roller skates, posed with a burger-and-shakes drive-in tray and looking straight at the "camera." Wearing thick glasses and a pencil behind her ear, the wrinkled Anna appears on roller skates in tight slacks, with a ticket-pad tucked seductively (ridiculously?) into her belt. Anna smiles slightly, with her well-manicured but wrinkled hand relaxing on her hip as a counterweight against the drive-in tray in her left hand and the change maker on her left hip. She is "serving" a car whose backseat occupant is a prepubescent boy (Generation Y?). He leans out the window, milkshake in hand, laughing hysterically in his youth league soccer shirt at Anna. The headline asks: "WILL YOU *EVER* BE ABLE TO RETIRE? WITH STOCKS PLUMMETING AND CORPORATIONS IN DISARRAY, AMERICANS' FINANCIAL FUTURES ARE IN PERIL. HERE'S HOW TO MAKE THE BEST OF IT."

Inside, the Notebook: Milestones section of the magazine featured a short story under the headline: "14 Years Ago in *Time*." The story was accompanied by a copy of the 1988 cover entitled "And now for THE FUN YEARS! Americans are living longer and enjoying it more—but who will foot the bill?" The cover photo shows a white man and woman who look to be in their early 70s, dressed in casual sweat suits and standing against a blue-sky-covered ocean. The couple is grinning broadly, and the woman pumps her fists into the air. The story suggests that the issue's theme was "new opportunities" for America's aging citizens.

It is not simply the news magazine that has discovered the value of acknowledging an aging readership. American *Vogue* magazine has produced a yearly issue devoted to aging and fashion since 2001. It might not be what you think. Thirty-three-year-old Jennifer Aniston was the August 2002 cover girl for "The Age Issue: What to wear, where to shop, when to lie, how to have sexy legs (and a sex life) forever, plus style setters from 16 to 80+." Similarly, in August 2001, model Amber Valetta, a young blonde with apparently flawless skin, appeared on the cover of "The Age Issue: What to Wear from 17 to 70."

Although a few models and actresses over age 50 appear in photographs that accompany the *Vogue* cover stories, most pictures are of younger women. It is also useful to point out that the older workers in these photographs have benefited from such "prosthetics" as cosmetic surgery, professional cosmetology, fashion photography, and other modes of glamour management. It is also significant that the balance of the photos in the magazines—most of them in advertisements—are of younger models. *Vogue* might be *saying* to Baby Boomer women that the magazine is relevant to them through its once-a-year photo essays that show us the well-preserved likes of former supermodel Lauren Hutton and the slinky designer Vera Wang, but it is *showing* them that they must work harder to defy their age.

The ads for such products as Botox Cosmetic and Neutrogena Visibly Younger hand cream underscore for Boomer women the need to engage advanced technologies to keep their suppleness and relevance. An ad for Botox in *Vogue's* August 2002 issue showed a smooth-faced woman with the tiny words "Your toughest wrinkle" printed across her forehead. The ad promised: "It took forty years to get it. And ten minutes to do something

Table 7.1 Elders in Technology Advertising

GENRE	INCIDENCE OF ADS	TYPES OF ADS	POSITIONING ELDERS
Newsmagazines: *Time, U.S. News & World Report, Newsweek*	0–2 per issue	Investment firms, biotechnology as investment, hearing aid batteries	Elder workers can adopt technology but they can appear as fish out of water with it.
Entertainment: *Rolling Stone, Entertainment Weekly, People Weekly, Biography, Premiere*	0–2 per issue	Gaming devices, Zip disks, stereo speakers, microwave oven, MasterCard, genealogy software	Younger-skewing magazines portray comical elders and hip Baby Boomers holding onto youth, and the youngest don't show aging at all. Older-skewing magazine celebrates the joy of aging actively.
Men's: *Playboy, Penthouse, Maxim, Esquire, Men's Health, British GQ*	0–1 per issue	Wristwatches designed for the high-tech set	Lots of technology ads, but they don't show gray hair. Gray-haired men, when shown, are depicted with other "power tools," such as expensive liquor and glamorous young women.
Women's: *Vogue, Glamour, O: Oprah*	0	—	Irrelevance.
Women's home: *Better Homes & Gardens, Martha Stewart Living*	1 per issue	AARP, MasterCard	Celebrates the joy and challenge of aging actively.
Ethnic: *Ebony, Hispanic*	0–2 per issue	Insurance investment firm, corporate sponsorship of cultural event	The aging worker is acceptable as long as the setting is wealth or glamour ("successful aging").

Business: *Inc.*, *Business 2.0*, *Fast Company*, *Fortune*, *Business Week*, *Kiplinger's Personal Finance*	0–3 per issue	Web site for entrepreneurs, wireless network provider, online broker, technology investment firm, multimedia projector maker, digital camera and laptop maker, Apple computers	Scant mention of elders in younger-skewing magazines (like *Inc.* and *Fast Company*) but more ads depicting older workers (CEO types) in more conservative magazines (like *Fortune*). Conclusion: Older workers are either irrelevant, or they are powerful decision makers.
Luxury: *Robb Report's Showcase*, *Architectural Digest*	0–1 per issue	Investment firm	Technology not typically paired with aging (which is more often seen in a healthy retirement setting). In the one ad where aging and technology are matched, a CEO type is seated, a ring of standing Gen Xers clad in black surrounding him. The point: Elders need to manage their wealth by staying on top of trends.
Technology: *MacWorld*, *Red Herring*, *Wired*	1 per issue	Computer wholesaler, data storage provider, Web hosting company	Decisive, discriminating elder men. No women.

about it." Botox is instant, it is scientific, it is a "non-surgical procedure." In another ad, in the 2001 aging issue, Olay's new "Total Effects" boasts a "Vitamin Lipid System" to "wash away the years." In another ad in the same issue, Nutrifusion offers "custom formulas" and "sensory matching" to moisturize the skin and "minimize fine lines." The fashion photos in *Vogue* magazine tantalize women with designer clothing with prices in five figures, but it is the ads that sell them the real merchandise. Along with lipsticks and nail polishes, the thousands of Boomer women who buy *Vogue* are also buying wrinkle creams and hair dyes—products that require ever-more-sophisticated messages of biotechnological success. It is part of the job.

Beauty and fashion magazines such as *Vogue*, of course, are unconcerned with the world of work, except for the maintenance of feminine beauty, including youthfulness.[2] *Modern Maturity*, on the other hand, has grown increasingly interested in work-related topics for aging readers, and technology is a common theme in the articles and advertisements that fall under this topic. In five bimonthly issues between fall 2001 and spring 2002, *Modern Maturity* carried nine prominent advertisements (quarter-page to double-page-spread) marketing products and services for older people who are interested in using technology. All of these products implied a leisure-related use of the technology. The ads included the Senior Explorer Internet service provider, featuring a band of smiling multicultural seniors gathered around a computer; the email PostBox service, used by a comically youthful pair of Glam seniors; the I-opener email service, also for senior surfers, hawked by a hip elder in cropped jeans; the Transcast television Internet service; Next50.com Internet service provider; and WebTV, which was heavily marketed to the senior surfer set. They also included ads for digital peripherals from Hewlitt-Packard—a digital camera and a printer. One ad was oriented more toward work, at least in the sense that I have argued that *Vogue* markets products for women who are engaged in the work of youth maintenance. L'Oreal's age-defying cream was pitched to women readers who wished to use a technologically advanced formula to fight the signs of aging.

Most prominently, six of the nine ads portrayed the life of the elder retiree, and technology was hinted to be a fountain of youth in that arena. All six ads selling Internet- and email related services and products featured elders who appeared to be either proud, relaxed, or extremely hip as they used or posed with the technologies. One woman sipped a cup of tea with her laptop at her side. One man reclined on a beach chair as he smiled at his keyboard, peering through bifocals. Another man, relaxing in his recliner with a bowl of popcorn at his side, smiled over his shoulder at us as he opened his email with a handy keyboard and his television set.

Several of the ads celebrated youthful activity, however, and they were the most intriguing to me because they departed from the more typical "golf haven retirement" portrayal of the senior found in *Modern Maturity*. The "e-mail PostBox" ad showed a woman who appeared to be about 60 years old standing with her arms wrapped around her neat little machine, as if it were a toy. She is shot from above, and she is peering upward at us, her legs and feet reduced to childlike proportions from this perspective. She is wearing a bright red, boatneck T-shirt, and her hair is reddish. The picture has been digitally altered so as to emit a "Photoshop" feel: She is standing on a bright yellow-orange background. One side of her body, including much of her hair, is tinted bright red, and her pant legs have been tinted lime green. The ad promises: "Never touched a computer? Now you can E-MAIL with the touch of a button. . . . No computers, no sweat." And the all-important mature marketing pitch is added in the corner of the page: "Save $20."

In an ad for Transcast, the subjects are similarly treated. The background is bright blue and orange, with a wavy swoosh separating the two color blocks. A couple on holiday is having the time of their lives. The man, at left, leans toward us in the foreground. He is wearing a yellow baseball cap, positioned backwards on his head, neon-blue framed wraparound sunglasses, a bright lime Hawaiian shirt, and a bright yellow and black camera for collecting souvenir photos. The woman, positioned behind him and to the right, poses with her hand on her hip and with a bright-yellow personal water-fan in her other hand. She is grinning broadly, wearing a broad-brimmed straw hat, replete with fake sunflower, a pair of gaudy white sunglasses, big earrings, a flowered swimsuit, and a decidedly immodest orange cover-up. Transcast uses the man's words to market its product: "psst . . . I met Rose on TV—all I did was send her a FREE email!" The reader is instructed how to order the product and, as with the PostBox ad, the emphasis is on ease of use and economy.

The elders who are addressed in these ads are presumably subject to more than just those appeals. They are reminded that they are missing out on a cultural artifact that has now apparently become important to everyone in the world except for them. Get with the program, Grandpa, they are told in everyday life. *Modern Maturity* lets them find a way to do so that is relatively painless.

Significantly, the products and services advertised in these pages have not become runaway success stories with the nation's elders. Instead, as computer operating systems become more accessible and hardware costs drop, older consumers are turning to personal computers and getting Internet accounts.[3] Advertisements in the magazine in 2002 signal an acceptance of this change: They market Iomega zip disks, laptop computers, and related items. The "simple surfin' senior" is becoming a relic of a brief past that quickly was replaced by the PC.

Geeks and Grannies

In the 108 advertisements I examined in thirty-four magazines published between August and December 2001, I found five main themes at work regarding the triple treatment of age, work, and technology. In the sections that follow I will talk about each of the themes in relation to one or two of the ads that typified it. The themes are:

1. Geek Mystique: Tables have turned
2. Pop Knows Better (than to ask a geezer)
3. Tool Time: Use it or lose it
4. Business to Granny: Customer boss
5. Active Aging: Keep the upper hand

Almost all of the ads I saw fell into one of these categories, and a few contained elements of multiple themes. Although I have independently named the categories and placed the ads within them, the advertising business obviously creates messages that can be clearly understood by target audiences. I feel certain that I have left little room for disagreement in matching ads to the themes named here, but I do not claim to have exhausted the themes that might be found in similar advertising.

Geek Mystique: Tables Have Turned

These ads typically exploit the "generation gap" between mature managers and tech-savvy Generation Xers to pitch a sale to the older managers. They are quite common in business magazines and are increasingly seen in consumer magazines. Generally, they use humor to make their point, which is some variation on the truism that what older workers know is outdated, and the wise ones among them have ceded at least some of their power over to the young knowledge workers who can help them stay afloat. The ads range from a tip of the hat to the young geeks to celebrating the outrageousness of this cultural-technological shift in business. Examples abound, including the following:

1. Black Rocket: Genuity, a Network Services Provider, marketed its Black Rocket eBusiness project platform with a campaign featuring a table crowded with white-haired members of a corporate board crowded around a casually dressed, young, grinning geek, who is standing proudly in front of the "Black Rocket" statue he presumably has built. The table is made of rich wood, round, with red leather chairs, and it is placed in the middle of a wood-paneled board room. The sixteen elders, all male except one and all wearing dark suits with white shirts, are performing a stadium "wave" in honor of the young man and his rocket. "You can do anything with a Black Rocket," the ad reads, then goes on to detail the service's basics

and how it can be obtained. The Black Rocket ad I saw appeared in *Business Week,* implying an audience of largely older readers. This "board of directors" was hyperbolically elderly, making the point that the project management tool need not be understood but beheld as a profit guarantor, a liberator from business uncertainty. The elders look at once ridiculous and victorious in their very uncorporate reaction to the geek. (They are, in this way, like the characters in the film *Cocoon,* who revel in their discovery of a means to eternal life.)

2. Nextel ran a high-visibility campaign advertising wireless business solutions with a series of photos, apparently hung in a corporate board room. The campaign ran in several business magazines. Each ad showed four portraits of aging directors from the imaginary company's past, with quotes from the directors shown coming from their mouths. In one such ad, three directors comment:

> "I hear we've got wireless access to the company network."
> "I hear we've got real-time financial data, from anywhere."
> "I hear they've got a machine that makes coffee."

The point is that the business world has changed tremendously since the "geezers" were in charge, but their company has remained successful by keeping up with the times. The target audience for the ad is meant to contact Nextel immediately so as not to risk being rendered obsolete with the aging directors on the corporate board room wall. *Call in the geeks before something bad happens.*

3. A third example is the campaign from Phoenix Wealth Management, which appeared in business magazines and high-end publications. In one of the Phoenix ads, a white man with graying temples, wearing a blue oxford shirt and striped tie, is seated confidently in a richly upholstered red chair. He is flanked by a band of four young men, all of whom appear to be in their 20s. They are dressed casually but hip, in various black outfits. Two are Asian, two are white. One of the white men sports an irreverently hip mustache and goatee. "Money. It's just not what it used to be," the ad advises. "The more things change in business, the more experience matters." The ad is an acknowledgement that the elder man cannot "survive" without the team of geeks who surround him, but neither can he "survive" without experience—his own business experience as well as that of his investment firm. The more unstable the world is, the harder you must look for stability, the reader learns. The upshot is: Let the geeks do their magic, we'll do ours.

4. An ad for Target suggests purchase of the Gameboy Advance through the Web site getintothegame.com. The ad I saw appeared

in *Rolling Stone* magazine, still heavily subscribed to by younger Baby Boomers and with a substantial audience of younger readers, although it has been losing ground in recent years to younger-skewing magazines, such as *Entertainment Weekly*. The Target ad for Advance shows a shepherd's flock grazing amid the low-production values of an Astroturf meadow (suggesting the graphics of a video game). Sheep surround an ancient sheep tender, with a flowing white beard and huge white eyebrows. Wearing biblical-era robes and sandals, he is seated on a rock in the field, his staff leaning on his shoulder as he is mesmerized by the handheld gaming device. In the background, the big, bad wolf walks away on two legs with a helpless lamb crooked under his arm. The (young?) wolf throws his head back in laughter, but the shepherd does not notice. The ad does not show an elder working with technology but an elder playing with technology instead of working. Its impossible treatment of aging suggests to the young gamer that the Advance is so cool, even this old geezer is addicted. Clearly, the old shepherd should be tending to his flock but he is powerless over the technology, which appears to be in league with the cruel wolf.

5. Finally, the ad from United Technologies Research Center is worth noting. In it, six men walk across a gray background, the blank area behind them splotched with a metallic "bloodstain." The two men in the lead are gray-haired, perhaps in their late 50s. The two men behind them are about 40, and the men behind them are perhaps in their 30s. The men in the rear are Asian. All the subjects are wearing black suits, dark sunglasses, white shirts, skinny black ties, and plastic pocket protectors. The headline is bold: "The punks who killed heavy metal." The copy below explains that the six are the scientists who developed metal foam, a light but strong material that will give the world lighter machinery at lower costs. A blend of old-style and new nerd, this band of subjects is part *Men in Black*, part *Reservoir Dogs*, part growth stock. It is humorous to see the elders leading the "band" of "punks." We are relieved to find that they are "really" scientists, and that heavy metal was only a play on words to conjure the bad-boy Generation Xer image. Placed as it is in business magazines, the ad assures us that these geeks are not representatives of some fly-by-night dot-com but can be trusted with our hard-earned investment dollars through the New York Stock Exchange.

The ads that represented the Geek Mystique theme were by far the most interesting of the lot for me to study. One reason for this is that they rely heavily on humor, suggesting an awkwardness and even nervousness about intergenerational function and dysfunction in the high-tech workplace. The ads project a ridiculousness about the Other from the perspectives of both older and younger generations portrayed within them.

Pop Knows Better (Than to Ask a Geezer)
Related to the Geezer Mystique ads, the Pop Knows Better ads do not necessarily imply that expertise belongs to younger folks, only that it does not reside within and that the discriminating consumer will find it. These ads are sometimes humorous but more often are direct and serious.

1. The most prominent of these ads appeared in the campaign by Charles Schwab securities. The ads contrasted the expertise of a Schwab investment specialist with that of another type of expert. Generally, the "other experts" were elder figures, a judge, a grandma, a family doctor. For example, in one ad, the headline states: "Both are seen as pillars of trust. Only one can assess the health of your portfolio." The photos showed a gray-haired man in a corduroy jacket with a stethoscope at his neck. He is identified as "Doctor Watson—delivered three generations of babies." The smiling Dr. Watson is leaning against a thirty-something blond woman in a dark suit with a string of pearls at her neck. She is identified as "Kathleen Hastings, Schwab Investment Specialist, McLean, Va." That Dr. Watson, as contrasted to Ms. Hastings, has neither a first name nor a hometown adds to his folksy appeal: He could be our family doctor; he could have delivered us. But would we trust him with our retirement plans? Doubtful. The younger Ms. Hastings, an expert in her field, is in McLean, Virginia, but we can reach her colleagues through the handy Web site that is prominently advertised at the bottom of the ad.
2. Macwarehouse.com, which bills itself as "Your #1 source for everything Mac!", produced a two-page ad on a bright yellow field with its bold blue and red logo lettering prominent at the bottom. The human figure in the ad is a bald man who appears to be in his 70s. He is standing with his hands in his suit pockets, looking upward at the fisheye lens, so his head is a distorted ball of wrinkled skin, 1960s conservative eyeglasses, and white-cropped facial hair. He is saying: "I want it. I need it. *Now what?*" Although the old man is slightly comical looking in his datedness, the ad lauds his business judgment: He is looking for the correct answer to his dilemma and presumably is sufficiently wise to act on it once informed.

Tool Time: Use It or Lose It

These ads, while not the most imaginative, are among the most plentiful of the treatments of age, work, and technology. They are ubiquitous in the backs of business magazines, and it seems impossible to page through an airline periodical without encountering three or four of them, at least. Typically, a gray-haired, white executive is photographed in a business setting, usually making some sort of presentation. Often, the product being sold is a presentation projector or related item, such as a laptop computer, lodgings at a hotel that caters to business users, or the services of a photocopying franchise. The advertising copy always assures the executive that he or she will be successful because of the product's capacity to provide organization, portability, and high-end effects at a sensible cost. Similar ads market goods and services that assist the executive in business at home, too, but the greater appeal of on-the-road-ware may stem in part from the practice of perusing business and travel magazines while in transit. I have learned from many interviews with older workers that they tend to take comfort from traveling with their presentation technology intact. By contrast, when they must learn a new application or plug in a new peripheral on the spot, it tends to make them nervous. It has been my impression that workers doing similar kinds of jobs in their 20s and 30s are much more relaxed about switching to technologies with which they have little or no experience: They approach technology intuitively and are less afraid than their elders of looking foolish when it does not immediately work. I believe that the presentation hardware marketing exploits this anxiety on the part of older business travelers who might like the assurance of bringing the comforts of home with them on the road.

Business to Granny: Customer Boss

The advertising world often hails its buyers through the vehicle of the buyer's own customer. The idea is to see one's business as others do so as to know how to profit. As America ages, we are beginning to see large numbers of ads that direct themselves to businesses whose customers are old or middle aged.

One such ad is from the United Parcel Services campaign focusing on "Moving at the Speed of Business." The copy tells the reader: "THEN you ran your company. NOW she does.... Meet your new boss. She's your customer." The ad goes on to assure businesses that UPS's global capacity delivers its customers the capability of being wherever and whenever they must be. UPS's familiar brown and yellow brand colors are prominent in the ad, which features a sepia-toned photograph of an old woman, who is at least in her 80s. She is standing, arms folded at her breast, in her front yard. We see the trappings of an old-fashioned working-class existence in the background. The woman is squinting and smiling. Her wrinkled face and arms are strange to see in an ad with such a closeup image. Her mouth is closed, but she apparently is toothless. The juxtaposition of this old woman's wrinkled image and the implication of her power serves the business owner notice that he must pay attention to his changing customer base and his organization's shifting balance of power. The idea that such a grandmotherly figure might make or break the business owner's enterprise must come as a shock, but the reader may relax, knowing that UPS's familiar brown trucks have everything well in hand.

Active Aging: Keep the Upper Hand

These ads tend to feature a single male subject demonstrating his expertise or satisfaction about his craft through the assistance of new technology. Women, though less common, are sometimes addressed through the subject choice, but the more frequent choice of men seems logical, considering the more frequent participation in the white-collar, paid work force by men 50 and over. The ads portray passionate experts who have wisely made carefully considered technology choices to help them in their pursuit of excellence. The men in these ads tend to suggest the artist, explorer, or other romantic element in our older selves. Technology is a background issue, a means to an achievement. Still, it is not a matter to be taken lightly, because only the best will do. We can find these ads in all sorts of magazines to which older men might subscribe (or older women, in some cases). Examples include the following:

1. Iomega advertised its CD-RW drives, Zip drives and disks, external hard drives, and network systems in an ad that relates a vintner's "passion" for fine wine production to Iomega's "passion" for protecting, storing, and sharing data. The man in the ad stands in a darkened wine cellar. His silver hair is among the lightest areas of the picture. He peers through trendy eyeglasses at a balloon glass containing red wine, which he examines with the passion of an artist. Underneath the photo of the wine cellar are smaller photos of the Iomega hardware that enable the vintner to approach his craft with such concentration. We are told that, as with the vintner, "Iomega takes care of your growing digital life, so you can take care of the rest of it." This man is a professional, an artist. That his life has proliferated with such complicated digital matters is an issue on which he should not have to focus, especially at his age when he has earned great prestige. Iomega will assist.

2. Sony ran a similar ad, which depicted a small-business owner, a porcelain restorer with a shop full of bathtubs and sinks, relying on the Cyber-shot digital camera and Sony Memory Stick organizing

system, together with the Sony notebook computer, to keep track of his business projects and communicate with suppliers and customers. Like the Iomega pitch, the Sony ad promises order with no fuss: "Ten hours a day restoring porcelain. One hour a week restoring order," so "you can get back to doing what you love." The hardware is laid out in a neat arrangement on the floor of the shop. A photo of an antique tub shows on the screen of the laptop, and the gray-haired craftsman peers confidently through his glasses downward at the image on the back of the digital camera in his hands. Soft light from the shop doors washes the shop in a beautiful glaze. Old, familiar things are everywhere, with the sleek, black, blue, and silver technology perched helpfully at the front. Sony will assist.

Regulating Bodies through Advertising

In writing about the film *Pumping Iron II*, Chris Holmlund emphasizes that muscularity by the female body-building contestants and judges in the movie is a concept that must stand in relation to that of "woman."[4] Only if we can read the bodies as still feminine is it acceptable to enjoy their muscular beauty, Holmlund contends: "The film makes their questions its own, marshaling images and sounds to ask: Is a woman still a woman if she looks like a man? Where is the vanishing point?"[5]

Another way of asking Holmlund's question is this: "Is a worker still a worker if she looks old? Where is the vanishing point?" I am arguing that advertising constrains the discourse on the aging worker. Content producers marketing goods and services to older buyers are careful to restrict images of aging workers to either white-collar "in-charge" types, whose power transcends age, or middle-class "carefree" types, whose proper place is implied to be outside the formal structures of paid work and restricted to a kind of leisure. These advertising themes use the regulating trope of "successful aging" to construct limits around what elders might do with technology.[6] It is either a tool for maintaining power over one's workers, in the case of the white-collar images, or it is a tool for the management of one's endeavor to age actively. Crucially, in the first case, advertisements rarely display images of older workers using technology themselves. Rather, these white-collared elders are hailed as decision makers with regard to the implementation of technology into their firms. In the second case, that of active leisure, the older "workers" almost always are shown as technology users: The product or service is depicted as an interface between the elder's passive, failed, obsolete existence and the anti-ageist vitality that the customer desires. By either measure, the marketer offers to help the customer to avoid elderliness. Executives are shown looking powerful, but they are not working.

Other elders are shown "working" with technology, but they are not shown as workers.

Again: "Is a woman still a worker if she looks old? Where is the vanishing point?" Although elder women appear frequently in advertisements that position them as leisure-prone customers, women in their 50s and beyond are still rare in ads that depict technology in a structured (paid) work environment. Junior executives in such ads commonly represent both sexes and different races, but senior executives almost exclusively are male. This, of course, is not much different from "real life," where top positions are held by women in only a tiny percentage of cases.[7] Still, it is useful to ask why, in a world of mediated images where almost nothing else is based on statistical truth, older women are ignored as potential images linked with technological power.

Elizabeth Grosz, in theorizing what she calls a corporeal feminism, argues that the body is understandable not just in and of itself but also in its relations to other bodies of people and objects. In understanding the body relatively, we in turn understand the practices in which the body engages:

> The body does not hide or reveal an otherwise unrepresented latency or depth but is a set of operational linkages and connections with other things, other bodies. The body is not simply a sign to be read, a symptom to be deciphered, but also is a force to be reckoned with. The energetics, or rather the politics of signification must here be recognized. The ways in which (fragments of) bodies come together produce...a machine.... In itself, the body is not a machine, but in its active relations to other social practices, entities, and events, it forms machinic connections.[8]

In knowing that the elder body's experience is shaped not only by its interactions with the young in a youth-conscious society but also by the physical, cultural, and financial capital within a person's reach, we are able to know something about how older bodies are expected to behave. That means especially what behaviors they are expected to refrain from, such as vigorous productivity and reproductivity. Sexuality, especially, must be avoided or, at best, hidden.

Perhaps the most striking illustration from the magazine ads I studied was the UPS ad that depicted an old woman as "boss" of the firm through her rising status as "customer." The power of the ad lay in its ironic, almost comical treatment of the old, working-class woman as a powerful body. Shot in closeup, she displayed deeply wrinkled skin, and her dress conveyed a lack of concern about the recording of her appearance as a signifier of status. She is behaving as an old woman should—at home, far from the glamorous world of corporate work—and the irony works because we know that her control over "the business" is tenuous, indirect, tiny, but real. She is practically the last person we would imagine to be "in charge." What we do not see among the technology-related advertisements are pictures of old women with leathery bodies seated at the controls of power: the board room, the presentation room, the corner office. How such bodies are supposed to behave in relation to these environments and the machinery therein is simply to remain absent. The implied lived experience of the wrinkled, toothless, old body is one of distance from the places where the action occurs.

Aging Workers on a Continuum

Kathleen Woodward has observed that it is more helpful to look at the different generations as a continuum of age rather than, as is more popularly the case, distinct and often oppositional stages.[9] In connecting human existence through the ages, we can build a scholarship that is less hostile from the influences of age politics. Can Woodward's observation about the need to construct a more supportive intergenerational scholarship be reconsidered for thinking about other potential continua along which all aged people might find themselves in relation to one another? For example, it might be useful to think about the aged worker along the continuum of power and control.

The white-haired executive, depicted in one of the two most common types of advertising about technology, is placed at one end of the continuum. He (or possibly she) "has it all," at least materially, and that truth bears enormously on how he experiences his life as an aging worker. The idealized middle-class retiree, who might purchase a computer in order to keep in touch with grandkids or go online to book a cruise, is at a more centered point on the continuum. He or she has "achieved" active aging and the opportunities it affords through a healthy and comfortable retirement. Still, it is possible that such an elder has had access to certain levels of power closed off. As a retiree or someone who at least is focusing on leisure in lieu of paid work, he or she has given up "worker" status. The technology that is being pitched to this worker focuses on rechanneling the desire to be a worker into the medium of successful retirement. Whether or not retirement is desired, it seems unlikely that this worker might have the choice of continuing to reach higher levels of power.

Fictive and experiential possibilities fill out the continuum of "aging worker" and inform particular encounters with technology, but it is significant to note that the opposite end of the continuum from the executive is not addressed through technology advertising. Whom might we find there? The unpaid or barely paid elder worker (often female, often of color), who might have little or no access to technology—and thus no need for advertising that sells technology? That is one possibility, which we are able to see easily through the filter of social class. But there are others. If *Fortune* is read by the CEO, who sees his mirror image in the ads that market network design solutions, what images are his wife seeing? Magazines that are popular among affluent older women—*Coastal Living*, for example—have little to do with technology and everything to do with the consumption of leisure through interior design, travel, and the like. This idealized elder woman of the country club set might have her own array of personal technologies—Subzero refrigerator, photograph-shooting cell phone, climate-controlled Lexus—but she is unlikely to have much control over how the rest of the world adopts and uses technology. On the continuum, she is at the opposite end of her husband, separated in the middle by a range of workers who might be, in fact, far less affluent but who on some level are able to exercise more power over their work with technology, especially in its interface with others.

Rather than life stage, people are placed along positions of power relative to one another. Gender, class, race, ethnicity, sexuality—all of these cultural markers influence to some degree where we might wind up on the continuum at old age, though none of them conclusively so. In its strong tie to materiality, the discursive tactics of advertising comment on the place where we are—and the place where we might like to be, as suggested by the power structures to which we adhere. Fortunately for us, the sellers assure us, consumer culture is standing by to help us reposition ourselves, as it is when we consider other continua, such as gender. In actuality, one's place on the continuum is not an indicator of progress, except in a material sense. Can we definitively assert that the harried executive is really better off than his wife, or vice versa—or that either is better off than the Sunbelt retiree who is busily emailing her niece on the bus back from the casino?[10]

In their fairly stable discourses on the intersections of age, gender, ethnicity, and social class, American magazines contribute to what Michel Foucault has identified as the political technology of the body.[11] Images in magazine editorial and advertising content are participants in the construction—renewed and repainted as norms slightly shift—of regulating discourses that help to bring about the body as we know it and experience it in the world. As with much media content (television and cinema come first to mind), advertising's images furnish the "knowledges" about the rules and limitations that have been set around the body by the forces of power. In turn, these knowledges might be resisted, or, what is more likely, accommodated through self-regulatory behavior. Advertising's volume and persistence bear testimony to the tendency for such "knowledges" to find resonance in the subject and result in confirming self-regulation rather than the option of counter possibilities. My observation of advertising depicting older workers and technology in American magazines in fall 2001 suggests not only that the spectre of the aging body is invoked around particular kinds of body knowledge, but also that the very "absence" of some types of image suggests further limiting knowledge. The conclusion is unsurprising: The affluent white male is the elder who may exercise power with respect to information and communication technologies, and it is the middle-class elder of either gender who may be consoled with the freedoms that information and communication technologies can afford.

From consumer advertising to social gerontology, regulating discourses of old age have suggested that the old body has experienced a loss due to decline. Emmanuelle Tulle-Winton discusses the "mask" of old age, first as it is defined by Mike Featherstone: senescence and physical decline "masking" the self that is inside the declining body as experienced socially; second, as Woodward notes, as the practices of "masquerade," the cosmetic surgeries, careful makeup ritual, and other means by which the body is disguised in order to avoid marginalization.[12] Old age as loss, covering the real self as a mask, and old age disguised as extended middle age, Tulle-Winton points out, are social means by which disciplinary practices bear on corporeal experience. A phenomenology of aging may be achieved where these social practices encounter resistance, through the transgression of aesthetic boundaries by the aged who refuse to be silently marginalized, she goes on to suggest. She adds:

> By putting forward the claim for a phenomenological approach to the old body, I am not simply suggesting that we make room in our understanding of old age for descriptions of bodily processes. Rather, we should develop ways of articulating the intimate ways in which we engage with bodily change, as an individual and also as a cultural strategy.[13]

Tulle-Winton's call for articulation of ways of bodily change brings me back yet again to the UPS advertisement. I commented earlier on its use as a vehicle of ironic comedy. Now I would like to suggest something different, something more hopeful. In all the magazines I looked at, I spotted exactly one woman of postmenopausal age who was portrayed as a business executive. She was Muriel Siebert, age 71, the first woman to have owned a seat on the New York Stock Exchange. Most of the male executives of the same age appeared to be in their 60s and 70s. What was entirely absent from these ads—and the ads about "the good life" of the surfin' senior—was any mention at all of the very old person, whose image graced the UPS ad. As Featherstone and Wernick have suggested, the *Modern Maturity* type of advertising, focusing on the healthy, active, ever-middle-aged senior, portrays as its primary goal the avoidance of old age through perpetual mid-life.[14] The insistent marketing vehicles of gray-haired 60ish executives and always-middle-aged senior couples furnish a regulatory discourse that disallows the lived experience of old age, pushing it ever to the margins of illness, absence, and death. The wrinkles and hollowed mouth of the UPS customer/boss, standing in her forever-front-yard, remind us that old age can, and should, be brought out of the shadows and that it is a potential moment of agency. If popular media images start to include more of the visceral accuracy of the aging (and often female) body, we can then, as a society, begin to talk more about the "project" of the self as a potential locus of resistance of external regulation and disempowerment. It becomes, at least potentially, possible to turn us toward tasks that lie away from the "body work" of forestalling elderliness.[15] A body that has lived a life need not be seen as a body of loss.

Leisure Workers and the Partnership of Marriage

Before closing, I would like to comment on the ads that position retirees as "leisure workers" through their adoption of information and communication technologies. Although we do know that the surfin' senior stereotype does have plenty of data to back it up, it is interesting that such a stereotype so powerfully outweighs that of another popular notion: the elder who uses technology to escape the confines of the disobedient body. In a special issue of *Images of Aging in Media and Marketing*, guest editor Maria D. Vesperi calls attention to both potential target audiences. She cites Dick Hebdige's essay on "woofies," well-off older folk, who are the target for the surfin' senior advertising vehicles that commonly market technology products and services in *Modern Maturity* and related magazines. She also mentions other essays, by Jacob Climo and Lee Leavengood, that suggest that restricted mobility and distance from children and other relatives are obstacles overcome by vulnerable elders who enjoy their ability to be "age free" on the Internet.[16] Such conclusions as the latter—freedom from corporeal aging—are cast into doubt by recent scholarly critiques that have concluded that the body is an always-present factor in the experience of an online life, and that life apart from the Internet cannot be so easily dismissed.[17]

Many, though not all, of such advertisements appear in *Modern Maturity* magazine, targeting an audience of people who are occupied with repositioning their postwork identities. Often, these ads picture a presumably married couple as a joint subject/customer, as do many of the leisure- and wellness-oriented advertising that is similar in thematic content. The idea is that the couple have entered a wonderful, liberating phase of life, and they are approaching this as partners in adventure.

In contrast, other advertising found in *Modern Maturity* addresses a single subject/customer and is often gender specific. Technology generally is marketed across gender lines, encouraging us to understand the domestic practices surrounding it as a site of new negotiations for the aging couple.

While we know that some computing functions are shared by older men and women (email) and some tend to be split (genealogy and greeting cards more for women, Internet financial study more for men), the sellers of ICTs recognize that a generation of retirees in their 60s and early 70s is charting the new place for unprecedented media/appliances in its homes. These homes heretofore have been highly demarcated by gender function: kitchen appliances, lawn appliances, even VCRs and remote control devices.[18] But the new ICTs are instruments that both aging women and aging men will use to work out their evolving identities, including, for some, new forms of paid work and work searches. They are both digital prostheses for these newly "homebound" couples and regulatory instruments of domestic power. In some cases, the regulatory instruments have shifted the balance of power in these "mature" marriages. In others, they have maintained it. In either case, it has been the creative or sustaining force of the people involved in the marriages that have made the difference. For instance, among several couples I have talked with over the past four years, I have learned that whether one or both "retired" spouses is using the home computer for some form of paid work can be a critical factor in the exercise of domestic power through technological control. Second, the use of the computer for unpaid or volunteer work might also make a difference. In the absence of either, the couples I have met most frequently report that their acquisition of a home computer has sparked an egalitarian trend in their partnership, which has been encouraged by other newly shared resources.[19] As Marsha Cassidy has noted, "By crossing the home's threshold, the computer becomes a 'disturbance' that jolts gendered domestic routines and spaces."[20]

The entry of ICTs into the domestic environment of the retired heterosexual couple might mean loss as well as gain for both parties. Women lose, for the first time in many cases, their dominant position over the interpersonal communication media in their homes. In doing so, they also lose freedom with a physical place in the home that almost certainly had been under their purview, such as the second bedroom ("sewing room"). As with much of the rest of what men's retirement has brought traditional homemakers, wives give up control. Women gain, on the other hand, a partner in responsibility for communication with the outside world. Men may lose dominance over the home's electronic media as the role of the computer grows more important in the home. In doing so, they gain a site of negotiation with their wives over how the couple will agree to present their joint and separate selves to the outside world.

Because the computer is a shared appliance that does not function as a shared medium, they must take turns, formulate joint statements, and, in the most communicative relationships, forge pathways together for gathering information and performing assigned or self-generated work. Many men and women have separately told me that this was a positive event in their marriages, though not always. Generally, men of this "Mature" generation have looked to the new home computer as a work surrogate, a means for reclaiming their career selves. This has been less true for women of this older generation. In any case, by the time most Baby Boomers retire, they almost certainly will have used computers in their work environments previously and will already have a home computer, perhaps a third- or fourth-generation machine. Additionally, women are likely to have worked outside the home and will share the foray into retirement with their husbands, if they are married. (Boomers are much more likely to be divorced or never married at all compared to "Matures.") Finally, retirement for Baby Boomers is unlikely to be as discrete an event as it was for their parents. New careers, step-down retirement, and other continuations of work will mean reliance on the home computer for paid work in many if not most Baby Boomer homes in the decades to come. Advertising of ICTs for this generation of retirees will no doubt have a much sharper work-oriented edge.

Although some magazines are geared specifically toward younger audiences, magazines that are less age dependent have awakened to what has popularly become known as the "tsunami" of the Baby Boom. In doing so, magazines targeting such general interests as homemaking and news are expanding their pages with special advertising sections focused on aging. Additionally, new Boomer-targeted magazines such as AARP's *My Generation*, for members 50–55, most of whom are still employed, and Meredith Publications' *More*, for women over 40, were gaining ground quickly in the 1990s.[21] Still other magazines, aiming for the generation older than Boomers, included *Mature Outlook*, *New Choices*, and AARP's *Modern Maturity*. Each of these publications had been experiencing major growth in technology advertising, especially *My Generation* and *Modern Maturity*, until the economic downturn in the early 2000s worsened, resulting in such cutbacks as AARP's decision to merge *My Generation* and *Modern Maturity*.[22] Even among more traditional aging-related products and services, advertisers are making efforts to "destigmatize" their acquisition by graying Boomers, as with ads for hearing aids for the "fashion-conscious," which focus on the currency of the technology being used in the product.[23] The "gray market" was still a relatively radical concept to advertisers a decade ago. Today, relatively few sellers of technology can afford to ignore it.

Advertising is a constant seesaw between stereotype and the new, new thing. In magazines—perhaps the most visually driven medium for presentation of this seesaw effect—we have seen the emergence of a new cast of elders in the service of hailing consumers. As argued, this emergence has mostly been propelled by the growing sector of older individuals with money who wish to spend, or may be persuaded to do so. Also significant, however, is the advertisers' consistent regard for the niche nature of magazine readership. Only if a magazine has a large segment of older readers (for example, business and home magazines) do we see a high count of ads that involve the triple theme of aging, work and technology. In magazines that older Baby Boomers and Matures are not expected to read, such as *Playboy* or *Vogue*, we rarely see references to elders in the ads at all. And, because the latter magazines are often connected to marketing the technologies of youthful desire—from lip gloss to backpacks—we might easily forget that elders of any type might be either subjects or objects of desire.

The landscape is changing, of course, as age-conscious Boomers continue to keep up their subscriptions to *Newsweek, Better Homes and Gardens,* and *Sports Illustrated.* Advertisements directed at those who might desire synthetic estrogen replacement, solutions for withering parents, and Cadillac Esplanades would not have been expected in mainstream magazines a few years ago, yet they are replacing pitches for Pampers and credit card companies these days.

Generally, the Boomers and their elders, the Matures, are being portrayed in complimentary ways in the ads. This is no surprise, because the older readers largely are being flattered toward a state of consumption in the process. What is encouraging is that they often are not being heralded for their stereotypical qualities, as in advertisements that say essentially: "You've worked hard. You've earned this nice thing." Instead, we see silver-haired sculptors, business people, and shrewd investors in the act of displaying their wisdom and work ethic. They are being lured by calls to purchase that involve the promise of efficiency, elegance, and precision, not simple exhalation. The appeal is the *love of work*, a career well spent.

As with most advertising, an extreme individualism accompanies these appeals to older consumers. Through their purchases, they are promised, they will be freed from the constraints that now encumber them, but absent is any suggestion that responsibility to address such encumbrances might be shared on a broader level. Money is freedom.

CHAPTER 8
How to Win Matures and Influence Boomers: Intergenerational Communication Through Self-Help

Tips for Age-Proofing Your Life

- Plan to live a very long life—80 or 90 years—and take steps now to guarantee the intellectual and social stimulation you'll want in your later years.
- Don't get trapped in yesterday's "linear" model of aging: Adjust your psychological, social, and financial expectations to support a "cyclic" life plan.
- *Envision new career goals and challenges. Intellectual flexibility and the ability to learn new skills and technologies will be key assets in a more longevous era.*
- Be prepared to reinvent yourself several times in adulthood—you may discover aspects of your potential you never knew existed.

—Ken Dychtwald
AgePower: How the 21st Century Will Be Ruled by the New Old[1]

In small and large bookstores a decade ago, it was a challenge to locate books on aging. Diligent browsers could locate treatments of retirement preparation, but little else. Betty Freidan's best-selling *The Fountain of Age* hit the shelves in 1993, preceding several similarly marketed volumes, such as Gail Sheehy's *New Passages,* Studs Terkel's *Coming of Age,* Hugh Downs's *Fifty to Forever,* and Mary Pipher's *Another Country,* among a flood of others.[2] Enter a Borders, Barnes and Noble, or good independent bookstore today, and you will find a different story. Many different stories.

Not only do works of fiction entertain us with coming-of-age tales, but nonfiction sections contain a growing number of aging-related texts. More specifically, publishers are demonstrating a growing awareness of the aging workforce and the impact that aging workers and technology are having on each other. Texts can be found even if some of the young personnel in the deskilled positions of chain bookstore clerks do not always know about their broad placement across the mega-stores. (To be fair, the chain stores do have some bright and well-informed staff. Because they employ large staffs compared with those of most independent booksellers, however, they are more likely to have some personnel who are not so well informed.)

While I was doing research for this chapter, I stopped in to browse the shelves of a large chain-owned bookstore in Columbus, Ohio. I was perusing the titles of the management section when a clerk, about 30 years old, asked if she could assist me. When she learned that I was looking for books that addressed aging and work, she shot me a patronizing smile. "You won't find books about older people there," she said, and promptly steered me to the Death and Dying section.

I imagined that independent booksellers and librarians might get a laugh out of that exchange, because it has been my observation that they generally have more than a passing knowledge of the contents of their inventories, not just the dominant organizational rubrics. That the clerk in Columbus could not understand why I wanted to keep combing the management shelves for books about older people was, I thought, a particular sign of the kind of structural lag that Matilda Riley has said accompanies changes in the experience of aging. As Riley has explained, changes in the human experience of aging have occurred much more rapidly in the current cultural moment than social schemes and norms have been able to adapt to.[3] Publishers know that aging is "hot" and are competing to provide significant works of nonfiction to market to readers who need to deal with the experiences surrounding the topic. But aging as a social phenomenon lags behind, sparking referents of nursing homes and Sun City snowbirds. For most people, the broad organizing rubrics have not been peeled away to accommodate the integration of aging into "everyday life" kinds of topics. Whether or not the margining off of aging—like that of gender, sexual orientation, and African American studies—is a good or bad thing to find in a bookstore is a complex question. For now, the least I can hope for is more Age Studies sections to appear in bookstores, expanding the imagery of old age as the new titles are doing, beyond the realm of Death and Dying.

The experience in the Columbus bookstore reminded me of the dozens of times when people have asked me what kind of book I have been working on. "It's about older workers and new technology," I would tell them. It always surprised me to see that what I meant as a simple answer puzzled the questioner. After a silence, people generally would muster the grace to continue the conversation: "My grandfather has an email account, and, man, you should see him surf the Internet." Or "I've been trying to talk my Mom into getting a computer, but she says she wouldn't use it." It seems that many people in their 20s, 30s, 40s, and 50s, when they think of older people as a class, relate first to their personal realm and not to their professional one. Of course, when we talk a bit more, they readily shift to their work experiences with older people, and I imagine that the chain bookstore clerk would have done so, too, if I had exercised sufficient patience with her.

My everyday interactions on the subject were different from my search engine experiences. Visit the Web sites of the major online booksellers and ask the "virtual clerk" about aging workers, and you will not be steered away from the business titles. On Amazon.com, a January 2003 search of keywords "aging" and "employment" yielded 61 results, and a search on Bn.com turned up 94 results (although about 10 percent had to do with work *with* the aged).

All of the books analyzed in this chapter were purchased in large or mid-size urban bookstores in the United States or from the Web site of Amazon or Barnes and Noble. Generally, the books clustered around two main themes: Intergenerational Crossings in the High-Tech Establishment (found mostly in the management sections in bookstores), and Our Aging World, (best-selling nonfiction sections). The first group focuses directly on workplace issues and highlights concerns about technology with remarkable internal consistency. The second, Our Aging World, is more diverse. Many of the selections concern implications for the world of work due to the forces of national and global demography; the upshot typically is that, with more people living longer, people are going to need to work longer. But other texts in the Aging World group point to more abstract approaches in considering the role of work in elders' lives.

Intergenerational Crossings in the High-Tech Establishment

An increasing number of volumes probe the multicultural workplace for managers, and some of these books examine the intergenerational workforce. None of the books assumes an audience of a particular age. Many engage in an "othering" of each demographic segment, as if the manager who might read the text might be ageless. Most of these books take the approach of selling multigenerational work teams as both a "fact" of modern life and a smart profit-seeking strategy. Some texts take a more legalistic approach, concentrating on pitfalls to avoid for the manager or business owner who does not want to be sued.

Most of these books do not take the issue of age segmentation head on but introduce it as one of several flavors of diversity. Such treatments almost always result in the "othering" of the much older and much younger employee, just as they introduce the worker of color, the immigrant, the minority religion affiliate, the woman, and the gay worker as "other." In general, by default, we read "normal" as white, male, Christian (at least in a secular realm), and somewhere between the ages of 35 and 45—characteristics of workers whose identities rarely seem to need explanation for the readership of such texts. The writers working in this genre, judging from their photographs and what they reveal about themselves in the texts and on the dust jackets, typically match this "default" readership, although a significant number of women work as co-authors in the field. The exception is the well-established "management guru," such as Peter Drucker, who is an older (though white) man and whose titles are enhanced by his name recognition. In such cases, the type size of the author's name is as dominant as any other element on the cover, including the book's title.

In books about "other" workers, the object is always to integrate the work group for maximum performance. For this reason, the decoding of diversity is a practical pursuit. We learn about the value system of a particular group, as defined through demography, so that we can, as managers, control its members' behavior for our mutual advantage. The workforce is aging; therefore, we need to understand how the Baby Boomers think about themselves so we can help them perform at higher levels of productivity and incur greater job satisfaction, making us, ultimately, more successful than we would have been without this given book. The texts never blur the lines of facile segmentations. We learn about Boomers and women and nonwhites (less commonly), but we never learn about the workplace experiences and expectations of older black women, for example.

A significant number of business-book authors are trying to help managers (the default position for whom is, of course, Baby Boomers) come to terms with the still-mystical Generation Xers in their midst. Although Generation Xers started appearing with their baccalaureate degrees in American workplaces as early as 1987, sufficient confusion about their identity and interplay with other generations remains to suggest a niche market for a considerable management literature that attempts to understand them and their place in business. Much of this head-scratching desire to understand the Gen X worker has to do with two notions, about which virtually all management books seem to agree:

- The Baby Boom and Mature cohorts make perfect sense to everyone: Matures, born before World War II and perhaps before to the Depression, came of age in hard times and built the expansion economy associated with the suburbs and other signifiers of prosperity. They carry their traditionalist and loyalist values with them proudly. Boomers, born between 1946 and 1964, were the Matures' children. They rebelled against tradition and had their perspectives formed by Woodstock, the Vietnam War, and Watergate. They brought us 1960s civil rights and '70s feminism. Then they grew up, moved to the suburbs, bought minivans and mutual funds, and took over.

- No one understands Generation X, the generation named after nothing. Doug Coupland, author of *Generation X,* suggested that we first understood his generation mostly through what it is not— not Baby Boomers, not Matures, not predictable.[4] Gen Xers are supposedly not loyal to any particular brand or logic. Before they were out of puberty, they had achieved a reputation as a band of slackers and skeptics. But, as virtually all the management literature points out, such stereotypes are grievous overstatements: Xers do believe in ideas whose principles they can understand and accept, and no company today can afford to be without their gifts of technological fluency and innovation.

Together, what these truisms tell us, through dozens of business books being marketed throughout the English-speaking world and beyond, is that older generations must be prized for their experience and trained for purposes of competence and retention, and younger workers must be cultivated for their amazing technological competencies, despite their lack of automatic loyalty for the company. Managers cannot afford to live in ignorance about the different groups' communication styles, desires, and goals. The so-called talent wars are resulting from an insufficient supply of today's best knowledge workers, the books remind us. Wise managers will win the wars by educating themselves about their most precious commodity, human resources.

Two tables accompanying this section detail how books look at this subject from two perspectives. In Table 8.1, I describe a cluster of books directed generally at older (Baby Boomers and beyond) managers, who must figure out how to work with a new generation of employees. In Table 8.2, I show examples of books that take on questions about the larger intergenerational workforce. The books in these two tables are listed here:

Young Einsteins
Petzinger Jr., Thomas. *The New Pioneers: The Men and Women Who Are Transforming the Workplace and the Marketplace.* New York: Touchstone, 1999.
Tapscott, Donald. *Growing Up Digital: The Rise of the Net Generation.* New York: McGraw-Hill, 1998.

Table 8.1 Managing Young Einsteins

Title	*The New Pioneers: The Men and Women Who Are Transforming the Workplace and the Marketplace* (1999)	*Growing Up Digital: The Rise of the Net Generation* (1998)	*Managing Generation X: How to Bring Out the Best in Young Talent* (2000)	*Managing Einsteins: Leading High-Tech Workers in the Digital Age* (2002)
Cover design	Hot pink, with multidirectional type wrapped around photo of X-tremely hip, 20 something, multicultural geeks, laptop included	Abstract graphic shows smiling, square bright faces. Back cover markets the companion Web site.	Gray and silver Xes form a repeat from the Generation X part of the title	Shot from above through a fisheye lens, geeky young white man surrounds his tilted head with hands shaped in Ls, framing the shot from the front of the camera. Back cover includes a sliver of a photo of the blue-eyed geek, cropped tight around his stylish glasses.
Dust jacket	"Shows that old-fashioned corporations are losing out to innovators engaged in creating collaborative workplaces, a value-added marketplace, and an economy overflowing with opportunity."	"Visit us online at... www.growingupdigital.com" "Makes a compelling distinction between the passive medium of television and the explosion of interactive digital media... shows how children, empowered by new technology, are taking the reins from their boomer parents...."	"The classic that exploded the slacker myth and introduced the world to the real Generation X."	"Frontline tips for managing the new economy's most valuable—and volatile—employees" "Proven methods, from alternative work schedules to on-site quiet rooms, for satisfying even the most hard-to-please Einsteins"

Author	"Author of Hard Landing" "Our navigator, the storyteller of the new opportunity economy"	"Author of the International Bestseller *The Digital Economy*"	"Studs Terkel for Generation X," an Xer himself	Pair of male business professors and consultants
Audience	"would-be pioneers who are considering a venture into the tumultuous world of small-business ownership"	Parents and teachers ("guidebook to kids' brains")	Managers	Managers of techies
Object of study	Gen X and Boomer entrepreneurs who visioned niches in new-tech new economy	Net-grown kids—N-Geners, born 1977–97 (bigger than Baby Boom at 30% U.S. population)	Generation X workers	New-economy techies, loyal first to their craft, then their tribe, finally their organization.
Who's who	Demographic forces—women and Gen X—changing the work environment. Baby Boomers in charge of mid- and senior management.	Raised on technololgy, N-Geners won't want to be stuck in companies that "don't get it." Boomer hegemony romanticizes Boomer-centrism and looks at new tools with mistrust	Managers: Boomers and Matures, used to rigid business models. Young talent: Gen X, raised on immediacy, information, independence, and (skepticism toward) institutions. Xers not disloyal but cautious.	Non-techies should be clear, supportive, communicative, and not try to be geeky. Techies (Einsteins, geeks) "come in all ages and genders" (but all the cartoons in the book depict young techies (mostly male) and older (non-techie bosses).

contd.

Table 8.1 *Continued*

What's technology	"IT is the friend of the middleman" by breaking up the mainframe and encouraging global infrastructure, empowering the individual business worker.	Video games make digital life second nature. Manual not required.	Elders mistake Xers' capacity to deal with large volumes of information fast as short attention spans, their powerful independence with technology for technoliteracy arrogance.	Technology generations divide geeks' tribes into "mainframe era graybeards, Unix Einsteins, the new PC-plus-Web generation," etc.
What's working	Self-taught coders like the Geek Squad, revolutionizing customer service by perfecting systems.	N-Geners' background in cooperation and diversity will produce a democratic business world.	Managers who coach but don't micromanage, give Xers cool tools and butt out of their work schedule and dress preferences.	Get geeks high-speed network and Internet access, cool tools, and explain why the work is necessary.
The world we live in	Conflicting: "air-conditioned sweatshops" of call centers and value-added marketplaces, influenced by innovative business radicals.	Kids have no "goof-off" time and are task oriented. They will bring this to the work structure. They will be able to organize "at the speed of light."	Employment isn't forever. Matures started downsizing. Don't blame Xers for protecting their careers.	Managers who want to keep their geeks must invest in training and prevent burnout. They need honest feedback and recognition but not micromanagement.

Table 8.2 Managing Generations

Title	*The Future of Work: The Promise of The New Digital Work Society* (2000)	*Generations at Work: Managing the Clash of Veterans, Boomers, Xers, and Nexters in Your Workplace* (2000)	*Geeks and Geezers: How Era, Values, and Defining Moments Shape Leaders* (2002)	*When Generations Collide: Who They Are, Why They Clash, How to Solve the Generational Puzzle at Work—Traditionalists, Baby Boomers, Generation Xers, Millennials* (2002)
Cover design	In-motion-look photo collage with binary code, globes, laptop, cell phone, stylized gears, year dates through 2012, and Web address: *www.futurework*	Head shots of ethnic 20-something male, 30ish white woman, 60ish white male wearing suit, and 40-plus black female.	Details from historic and contemporary graphics with images related to World War II and current America. A young, blond woman balances a portfolio and the *Wall Street Journal* on one arm, a baby on the other.	Slide puzzle squares in disarray. Effectively cuts pictures of white-collar workers into body parts: gray head, young Asian woman's eyes, turkey-neck of middle-aged woman wearing pearl necklace. Thirty-ish white male head appears almost in full. Back cover photo shows the squares properly arranged: The solved puzzle depicts a gray-haired white man leading a team of a middle-aged white woman; young, white male; and, in shadowy background, young Asian woman.

contd.

Table 8.2 *Continued*

Dust jacket	"How can you survive in a workplace that is rapidly changing because of revolutionary changes in technologies?"	"In our age-diverse workplace of conflicting work ethics, dissimilar values, and idiosyncratic styles...."	"Our youngest leaders matured in the glow of computer screens; our oldest in the shadow of the Depression and World War II."	"Bridge generation gaps at work by understanding the differences that drive generations apart."
Author	Dr. Charles Grantham, media guest, business expert, and academic.	Ron Zemke, Claire Raines, and Bob Filpczak, Mature, Boomer, and Xer, respectively: business trainers.	Warren Bennis, prolific management author, and Robert Thomas.	Lynne Lancaster (Boomer) and David Stillman (Xer), communication strategists.
Audience	Workers and managers: "You."	Managers of intergenerational work groups.	Good managers, aspiring leaders, future heroes.	Managers, employees, entrepreneurs, professionals.
Object of study	"How the workday is changing in response to new technologies, how the individual's sense of self-worth may be affected, and how employees and managers at all levels ... can work together to create new communies defined by heightened communication"	The "problem" of four generations of workers inhabiting the workforce together: Veterans, Baby Boomers, Gen Xers, and Nexters.	Understand conditions under which two distinct groups, under 35 and over 70, had their leadership potential honed. Identify critical shared qualities of both groups, including "youthful curiosity and zest for knowledge."	"Four distinct generations of employees, glaring at one another from across the conference table."

Who's who	Traditional workers—older; postwar 1950s values, demand longterm job security, define loyalty as tenure, believe employers responsible for career growth. Emerging workers—Gen X, delink security, commitment to job; loyalty is accomplishment; take personal responsibilty for career growth.	Veterans: formed by WWII values, "workplace hangers on," keepers of the grail, true traditionalists. Boomers: formed by Me Generation and Civil Rights values, workaholics, holding on to workplace power. Xers: deeply segmented, change masters, work to live (not live to work), smaller, more tech-literate group than Boomers. Nexters: Optimists, don't dislike their parents, achievement-oriented, raised on video games.	Geeks: Grown latchkey kids building adventurous careers in unsure times with the best technology available. Geezers: Traditional work ethic led them to build on conservation and responsibility; adjusting to shifting sands of today's workplace. They have become adept at anticipating and reacting to change.	Traditionalists: D-Day; head down, onward and upward, loyal, experienced, consistent. Boomers: Cold War, Civil Rights; it's all about me; work insanely hard; optimistic, formal feedback, documentation. Xers: parade of images; show me the money; skeptical; media saturated; informal feedback. Millennials: concern for personal safety; techno-savvy; practical; diversity; instant feedback.

contd.

Table 8.2 *Continued*

What's technology	"If you're over 40, [you remember] when information stored in computers was almost impossible to get." Late adopters of technology in the workplace becoming disempowered.	Veterans need traditional classroom training and will adopt technology readily to be able to make work flexible. Boomers gravitate to technology as a challenge, a way to make a difference. Xers expect access to decent technology and recognition of merit. Nexters know technology but need mentoring in other ways. They can mentor others about technology.	Geezers saw and participated in huge technological advances in postwar years. As leaders, many invested in bringing changes to the workplace and have continued to do so. Geeks: Expect to participate in technological development and innovation as part of their routine adventure. Some (Michael Dell) have acted entrepreneurially to effect such innovation in the sustainable environments they control.	"The technological revolution has exacerbated the situation, [putting] a greater divide between the generations who grew up with technology and the ones who are playing catch-up."
What's working	Internet enabling new forms of collaboration and community at work. Younger workers are driving the intensity of these experiences.	Flexible teams that respect diversity and follow clear communication models.	The authors' geeks and geezers designed own learning strategies for each stage of life. They build and maintain intergenerational work and social networks.	Get geeks high-speed network and Internet access, cool tools, and explain why the work is necessary.

| The world we live in | Generational shifts and pervasive communications technology pulling us into new collaborative work patterns. Design perspective on technology must replace engineering perspective for workplace of the future. | Workplaces are growing more, not less, complex, both demographically and technologically. Never assume the limit of a contributor's scope. Actively promote collaboration that is sensitive to each generation's needs (security/loyalty, accomplishment/recognition, professional development/contribution, and belongingness/learning). | All leaders encounter "crucibles" that can either break them or act as transformational moments in their personal and work lives. Whether one will become the first woman with a seat on the New York Stock Exchange, found E*trade in his late 50s, or found an influential video game review site at age 18 depends on that individual's response to life's crucibles. | "There's a talent war out there." Business are losing too many traditionalists "because they're feeling outmoded," too many Boomers to free agency, and too many Xers to "not feeling at home." Not enough Xer workers exist to take the places of retiring Traditionalists and Boomers. All generations must be trained and accommodated. |

Tulgan, Bruce. *Managing Generation X: How to Bring Out the Best in Young Talent.* New York: W.W. Norton, 2000.
Ivancevich, John M., and Thomas N. Duening. *Managing Einsteins: Leading High-Tech Workers in the Digital Age.* New York: McGraw-Hill, 2002.

Managing Generations
Grantham, Charles. *The Future of Work: The Promise of the New Digital Work Society.* New York: CommerceNet Press (McGraw-Hill), 2000.
Zemke, Ron, Claire Raines, and Bob Filipczak. *Generations at Work: Managing the Clash of Veterans, Boomers, Xers, and Nexters in Your Workplace.* New York: Amacom (American Management Association), 2000.
Bennis, Warren G., and Robert J. Thomas. *Geeks and Geezers: How Era, Values, and Defining Moments Shape Leaders.* Boston: Harvard Business School Press, 2002.
Lancaster, Lynne C., and David Stillman. *When Generations Collide: Who They Are, Why They Clash, How to Solve the Generational Puzzle at Work—Traditionalists, Baby Boomers, Generation Xers, Millennials.* New York: HarperBusiness, 2002.

Gen X: The Future is Now

What most interested me about the management books designed to explain Generation X to older executives and managers was the irony that people in their 40s, 50s, and beyond are paying upwards of $25 to have their children's or grandchildren's cultural identities explained to them for the purposes of business communication. Managers are told about Gen Xers as if these strange workers among them could well be Martians. They learn of the group's shared sociological characteristics: latchkey children of divorce, grew up playing Atari games, expect informal mentoring and feedback systems because they are uncomfortable with senseless bureaucracy. But almost every single characterization of these younger workers is in terms of the dilemmas they present: How can managers motivate them, how can co-workers collaborate with them? They are presented as competitors, interlopers, and, at best, puzzles. It is an astonishing irony that such literature, while propounding the value of intergenerational relationships for work, never asks the reader to imagine the Gen Xer as having a shared identity with young intimates in his or her life, such as sons and daughters.

Even so, the books that probe the identity of the Gen X worker do "sell" the elder reader on more positive treatment of the Gen Xer than the authors believe is commonly found. They recommend, generally, these steps toward success in the intergenerational workplace, with emphasis on making the best of employing younger workers:

1. If you ask Gen Xers to do something, explain why it is necessary and then offer feedback about how it helped the organization work toward its mission. Xers hate pointless work, and they view work whose purpose they don't understand as pointless.
2. Give Xers the time, space, and tools they need to do the job. Don't impose rules on them out of convention when they will see the rules as pointless. For example, don't require them to be in the office at 9 A.M. if they work better in the afternoons and evenings. Also, Xers will expect downtime—real weekends and vacation time. They will put in extra hours to complete a significant project, the management gurus say, but they will expect time off to go rock climbing as a reward.
3. Give Xers opportunities for professional development. More than salary increases, they value opportunities to build their careers, through learning significant technical applications or being assigned to influential work teams. They are loyal to their own professions rather than to their organization. This is something that is especially difficult for "The Organization Man" of the 1950s to understand.[5]

I found it both sad and ironic that a generation of managers who likely had significant personal relationships with Gen Xers need to be given such advice as explaining why an assignment is important when you give it to a Gen Xer. One has to wonder what kind of parents the leading-edge Baby Boomers are to the Gen Xers in their own lives if they have not already adopted such helpful communication guidelines. With the lines between work and the rest of life blurring at an escalating rate, it seems a surprising pretension that the management literature so completely cordons off questions of demography—generations, race, sexuality—as an office issue that bears no connection to "real life."[6]

Envisioning what the directory of a typical Boomer manager's email account must look like, I wondered if, like my own account, the trail of "office" messages might be sprinkled with the trappings of a life not so easily segmented: notifications from school groups about meetings, email exchanges with older children about evening plans, queries from a distant older relative about genealogical data, perhaps. Surely, the management gurus realize the messiness of real people's work lives. (Or, perhaps the unofficial, nonwork life is read through the dominant lens as not fitting into the still male-oriented, separatist life of the American manager.)

Among the ironies I drew from this observation was that the management books preach sensitivity to experiential difference: Older employees may desire flexible work schedules (including telecommuting arrangements) in order to care for elderly relatives and to visit younger relatives out of state. Women may value on-site day care facilities at the firm. Younger employees might give you a seventy-hour work week to get the important project finished but expect you will release them without question when they want to hike through Europe with a college buddy. Yet, among all these and other wise imperatives, the authors do not interpellate the manager as human, as experienced, as flawed, and as a person with a particular orientation toward work and technology that likely was distinct from that of their Gen X workers.

In some of the conversations I had with older managers, they acknowledged links between their Gen X employees' identities and those of their own family members. This acknowledgment, however, occurred much less often than not, and I observed that the managers were disinclined to see their fellow workers as members of a common generation apart from the workplace rather than as a cause of perplexity inside it. Stereotypes are much more facile than reality, and they come from all directions.

Boomers: It's All About Me

The how-to take on Baby Boomers, for managers, acknowledges that Boomers are better educated than Matures and more experienced with technology, and they feel an entitlement to receive opportunities in the high-tech workplace, yet they are left in the dust by Xers who cut their teeth on Java Script and who fail to be impressed by the Boomers' canned PowerPoint slideshows.

The main message of the management literature *vis à vis* Boomers is that, with the retirement of the Matures, Boomers have all the keys to the management kingdom (Bill Gates), except for those grabbed away by a few feisty Xers (à la Michael Dell). Not only are there plenty of Boomers, but they are aging in the business marketplace. They are losing jobs to downsizing, coming back as consultants and Web-meisters, and filing suit against sloppy employers who do not observe the Age Discrimination in Employment Act or Americans with Disabilities Act. Boomers, the literature suggests, are the new dinosaurs, but, like T-Rex, they are too large and mean to be ignored. They must be accommodated into the business ecosystem, or too few talented people will compete for the spoils of business war.

Generally, management books describe how to deal with Boomers in terms of these themes:

1. Place technology in the hands of Baby Boomers and challenge them to make a difference for the firm. Give them a reason to stay. Nothing appeals to them more than their own success, when they can see its results in action.

2. Baby Boomers tend to like technology, but they need to be able to see it as a way to "retool" for a better reality (either for themselves, the organization, or for society). They enjoy professional development opportunities that come with new technology.
3. Older Baby Boomers are looking for ways to overcome their long record of workaholism. Telework and flexible careers are attractive options for many of them, especially women. Early-retiring Boomers like to "retool" their careers by bringing their experiences back to work in less consuming ways. Most cannot afford to truly "retire," because they have not saved enough money, their private pension systems are inadequate or nonexistent, the stock market has lost too much of their savings, or they took on much more debt than did their parents' generation.
4. Baby Boomers like to believe they are "hip" about everything, including technology. They will quickly find that Generation X coworkers and employees judge them by their technical competence and sometimes little more.

Matures, Traditionalists, Silents—the Elders

Implicit in the texts of books devoted to managing older workers are two common conclusions: They can be difficult to deal with, and they can be even harder to keep. Virtually all the how-to-manage books acknowledge that maintaining a productive segment of mature workers is an essential business survival strategy for several reasons, including the comparatively small size of the Gen X replacement workforce, the expense of recruiting the best new talent, and the premium that has been placed on diversity in the flattened hierarchies of contemporary organizations. None of these books assumes older workers' right to expect fulfilling career opportunities in their later years, although management readers are instructed about using care not to discriminate against elders in a legal sense.

Instead, most of the descriptions of older workers deal with the obstacles they present to the workplace—the intergenerational communication problems and tensions that result from having them still around due to their increasing longevity. The books teach how to get the greatest productivity out of the elder, both through recognition of his or her special experiences and through a rigorous updating of skills. Most of the advice is about social and cultural distinctions brought to the workplace by the older employee, and almost no attention is paid to physical issues, such as potential disabilities and ergonomic concerns. The how-to books advise the following:

1. Recognize employees' record of contribution and demonstrate that loyalty is valued. Coach younger workers about interacting with respect toward older workers.

2. Help elders keep abreast of technology in a classroom setting. Traditional learning strategies are more appealing than informal ones.
3. Older employees and younger ones might mentor each other in key areas of strength: Older employees might share company history and customer service strategies; younger ones can help elders learn technology.
4. Older employees are starting to desire telework and flex-time so they can manage their changing personal lives more desirably. They can still make important contributions to the workplace if appropriately rewarded.

Popular literature on aging and the workplace is tremendously consistent. Differences designed to be distinctive are, essentially, small. Examples include cut-off-point years for generations (did the Baby Boom last from 1946–64 or 1945–60?) and nicknames for the generations (Traditionalists or Matures, Generation Y or Millennials). The primary points of most of the books were these: The Baby Boom is the largest generation ever, it is aging, and its interplay with other generations, especially in the workplace, cannot be overstated. This interplay is being influenced by the escalating role of technology, whose rapid changes are helping to locate the significance of younger generations in the workplace and potentially diminishing the contributions being made by their elders. Finally, the resocialization of the workspace into collaborative team enterprises, with obsolete hierarchical work relationships falling away, compound the need to emphasize the negotiation of generational communication and power relationships at work.

The intersection of age and gender surely contributes to the messages that come through popular discourse on technology, work, and age. The books I analyzed were written largely by men of various ages, although a few had female co-authors, and their audiences, while of mixed gender, were addressed in the dominant masculine style of business, which is disconnected from the realm of the personal. These writers portray technology as a set of tools to help managers accomplish their overall mission. Furthermore, they portray human resources as another set of tools that must be manipulated to work productively with technology. Such portrayals fail to acknowledge the messy encounters between technology and particular groups of workers, as distinguished by gender and age. Baby Boomer women, for example, have negotiated their own models for the installation of computers into the contested spaces of home. They might be more likely than their male counterparts to view "home-work" through technology as necessary engagement in the softer practices of human interactions that they feel are necessary to help them keep their "place" at work from the remote locus of the home. Boomer women who feel "guilty" about ducking out of the office to meet the demands of their private-sphere roles of wife and mother—and who already are adept at bringing "home" to their "work" environment—are likely to experience the creep of work into home through the technologies of the Internet, the modem, and other aspects of computing.[7]

The how-to books on managing the generations stops well short of interrogating the gendered means by which the home computer is being used in the transformation of the work force. For Traditionalist/Matures, white-collar male workers have always brought work home; home has been, for them, a place of leisure where they have underscored their dominance through the practice of conducting esteemed paid "home-work," which effectively excuses them from routine domestic engagements. For women of this generation, the home has never been a site of leisure but always the hub of domestic work. They likely entered the workplace later than their husbands and received less pay, and they continue to privilege their traditional dominant responsibilities while at home.[8]

It is in the life of the Baby Boomer woman that we find work most intrusive through the home computer. In part because of 1970s feminism and apparent economic necessity in this age of consumption, comparatively fewer women from this younger generation have viewed the realm of work as secondary to the home. While they remain primarily responsible in most cases for the routine work of running the household, they have added on the take-home work that is made possible through contemporary business computing practices. If you get email from a co-worker at midnight and again at 6 A.M., there's a fair chance the author is a Baby Boomer woman. For her, the work-home boundary blur has felt both empowering and suffocating.[9] Interestingly enough, the authors of the how-to-manage-generations books do hit on this Boomer home-work ethic, but in a surprising way. They discuss it only as background to explain Generation Xers to their older readers: Gen Xers demand real time off from work, the authors contend, because they grew up starved for leisure time with their workaholic Baby Boomer parents.

Because the management-advice books aim at a large audience, no attempt is made to describe the situatedness of the intergenerational workforce within industries or, more important, within firm size. As a result, authors dispense advice about getting training and benefits for older workers free of such concerns as the relative size of the organization's budget for such expenditures. At a time when two-thirds of all new jobs are being created in small businesses in the United States (firms with fewer than twenty workers[10]), such generic approaches to management cannot advance the concerns of older workers. As noted by Martin Sicker:

> [T]he shift in job creation from large to small and frequently new businesses has an anomalous consequence for older workers. Small new businesses tend to emerge in competitive industries characterized by rapid development and change, and therefore offer employment opportunities primarily to people with the necessary skills, having very little capacity to provide the retraining required by many older workers displaced from more traditional industries and occupations.[11]

The how-to-manage books discuss older employees largely in terms of rehabilitation and retention and only superficially in terms of recruitment. Most of the books have an obligatory section on how to attract the older worker, using such techniques as constructing a corporate Web site or video that tells the firm's story and champions its traditions. The books do little substantively to challenge the prevailing logic of employers to downsize or avoid hiring older workers for jobs with higher levels of technological skills due to presumed cost-ineffective retraining requirements. Among scholars to establish a link between age and job loss due to technological factors in such circumstances are New York University economists William J. Baumol and Edward N. Wolff, who have also demonstrated that older workers laid off in a technological area will be without replacement work for longer periods than other workers and are more likely to become permanently unemployed.[12] In contrast, the authors of the how-to-manage books keep the focus on the employment of older generations as a means of maintaining an organizational asset (human resources) by directing their moral centers at the interests of the audience/managers (and therefore the firm). In doing so, they seem to show sympathy for older workers' interests but are, in fact, able to bypass the realities such workers can face when jobs are lost or denied.

In this organization-centered discourse, some of the salient issues for employees never get mentioned. The stuff of real life, heavily colored by one's life stage, may deeply influence an employee's take on the work experience but does not "belong" in the masculine literature: menopause, depression, and divorce, for example. Apparently, a safe management is one that has memorized the buzzwords of diversity but keeps its distance.

Our Aging World: Is the Sky Falling?

In addition to the how-to-manage-generations books, popular literature has produced a large number of more general titles on the so-called longevity revolution. Many suggest that the aging of the Baby Boom, coupled with declining fertility rates inside and outside the United States, presents an impending national and global "crisis" in the coming decades. Typically, the books make suggestions on how to solve the crisis, by some combination of these efforts:

1. Extend the length of mandatory work before collecting Social Security.
2. Cut Social Security benefits (e.g., eliminate or scale back cost-of-living increases).
3. Privatize Social Security, in whole or in part.
4. Governments and/or employers should invest in lifelong learning to keep workers' technological skills current.
5. Workers should invest in their own lifelong learning and consider themselves mobile free agents.
6. Employers should make their organizations attractive to elders, who might like to work on part-time or short-term contracts in lieu of collecting Social Security benefits.
7. Entrepreneurs and politicians had better pay attention to the swelling new "gray market." Niche services, such as travel, health maintenance Web sites, and assistive technologies development, will be heavily influenced by this market. There will be no market for big, suburban houses or for mutual funds.
8. Baby Boomers had better expect a scaled-down lifestyle in their retirement years. They can't afford the one their parents have.

Overall, the message is an ominous one. One gets the picture of a self-absorbed generation careening toward maturity, scared to pull its blinders off. All of the books I looked at in this category were published prior to September 11, 2001, and, although they disagreed on the fine points, they all agreed that the aging of the Baby Boomers would strain the stock market. Now that the dot-com bust has preceded, and, in fact, re-routed the massive entrée of Boomers from the retirement circle, it is hard to imagine that there is much to cheer about on the economic front. Still, almost all of the authors of these books reference the Baby Boomers' stereotypical "optimism" and capacity for education. The tone in which the Boomers are discussed is *paedocratic*[13]: These are the culture's eternal children, who experimented in the 1960s, got married in the '80s, and finally discovered money. Perhaps, we are left to muse, as the bottom falls out of the economic future, the Boomers will prove to be a generation that at last can take care of itself.

Table 8.3 shows at a glance how four popular books on the "longevity revolution" from the late 1990s treat their subject. Here is a lengthier listing of titles:

> Smith, J. Walker, and Ann Clurman. *Rocking the Ages: The Yankelovich Report on Generational Marketing.* New York: HarperCollins Publishers, 1997.

Table 8.3 The Sky is Falling

	Title			
	The Fourth Turning: What the Cycles of History Tell Us About America's Next Rendezvous with Destiny (1997)	*Generations Apart: Xers Vs. Boomers Vs. The Elderly* (1997)	*Gray Dawn: How the Coming Age Wave Will Transform America—and the World* (1999)	*Agequake: Riding the Demographic Rollercoaster Shaking Business, Finance, and Our World* (1999)
Cover design	Silver and black waves flow from a segment of a small analog clock in the bottom right corner. Above the black-and-orange title graphic is the proclamation "National Bestseller."	Three headshots across front: Xer with multiple piercings sticks her tongue out for the camera, Boomer wearing conservative suit and haircut looks serious, elderly woman looks pensive. All are white.	Large, all-caps lettering screams the title. Interwoven down the cover is a sub-subtitle in red: "There's an iceberg dead ahead. It's called global aging, and it threatens to bankrupt the great powers. As the populations of the world's leading economies age and shrink, we will face unprecedented political, economic, and moral challenges. But we are woefully unprepared. Now is the time to ring the alarm bell…"	In faded blues and pinks, a photo of an hourglass tilts, with about half its sands already "lost" to the bottom half, half still to go.

Table 8.3 *Continued*

Dust jacket	Generational theory of history leaves reviewers with "mixture of terror and excitement."	"Essays on all three generations" but the focus is "What will be the impact on the younger generations when the Baby Boomers reach retirement age and join the elderly in demanding a huge slice of the economic pie?"	Washington insider luminaries blurb: Peggy Noonan (Reagan speechwriter), former Federal Reserve Board Chairman Paul Volcker, Reagan Secretary of State George Schultz, Sens. Sam Nunn and Warren Rudman, and news anchor Diane Sawyer. Sawyer writes: "Pete Peterson is the watchman on deck warning us all to wake up.... His facts are explosive."	"The new age of humanity will require wrenching adjustment to everything from property values to pension promises, from investment strategies to the world of work, from youth culture to your lifestyle."
Author	William Strauss and Neil Howe, generational experts.	Editors: Richard D. Thau and Jay S. Heflin, co-authors of *Get It Together By 30 and Be Set for the Rest of Your Life;* collected essays from authors of all three "generations"	Peter G. Peterson, 72, chairman of the U.S. Council on Foreign Relations and former secretary of commerce.	Paul Wallace, Cambridge-educated British writer and broadcaster of economic issues.
Audience	You: "whatever your stage of life."	Generation X.	"Business readers" and "those who care about public policy."	Broad: "The agequake is affecting you now. Your work, your money, even your love life are all subject to the shockwaves."

contd.

Table 8.3 Continued

Object of study	Generations in history, with emphasis on the current "third turning," an unraveling period when culture destabilizes, and the coming "fourth turning," which will bring a crisis to make way for a new stable period.	How three generations view each other and potential for disaster in the future.	Aging of America and the world—"a social, political, and economic time bomb."	Demographic changes (rising longevity and declining fertility) on an international scale, as they intersect with economic realms including the stock market, pension funds, property values, and ageism at work.
Who's who	GI generation: triumphed in the '30s and WWII crisis, heroes Silent Majority: enjoyed the '50s high, artists Boomers, "awakened" during '70s and '80s; prophets 13th Generation (roughly, Gen Xers), "struggling to adapt to our splintering world," nomads Millennials: (roughly, Gen Y), future heroes.	Elderly: Drawing mega-benefits but not as large a group as Boomers. Boomers: Feel entitled to the benefits of their parents. Xers: Can never expect as high a standard of living as Boomers yet expected to finance their parents' retirement.	Geezers: includes himself, growing number of elders leading the globe into crisis through reception of public pension benefits that young workers cannot finance. Young people and "future young people": Need policy solutions now or will require "Draconian measures later."	Boomers: "Methuselahs" will have to work longer because of anticipated bear markets. Northern and Anglo nations: Declining fertility will strain pension economies most. U.S.: Liberal immigration laws will keep it "young."

Table 8.3 *Continued*

What's technology	"Every Fourth Turning [crisis period in the cycle of generational time] has registered an upward ratchet in the Technology of Destruction."	Not an issue for these authors.	"High-tech Catch 22": Falling birth rates will mean smaller armies, greater demand for technology in military, yet how will U.S. afford it when senior electorate instead demands expenditures in high-tech medicine?	Technology mentioned only briefly: "The central problem for older workers is that the use-by date on their expertise is closing in all the time. In the past, skills and knowledge retained their value for quite long periods. Now they depreciate almost as quickly as computer equipment...."
What's working	Boomers: "The elder goal will not be to retire but to replenish or reflect or pray. The very concept of retirement will fade as Boomers pursue new late-life careers, often in high-prestige but low- (or non-) paying emeritus positions."	Disagreement on this: Some authors want benefit cuts, extended employment age. A few believe current policies are fine and elders deserve current levels of benefits.	Nothing. Needed: Benefit cuts, forced savings, longer work lives abetted by anti-ageist work policies, pro-child policies to swell the workforce, filial obligation policies.	Boomers, better educated than parents, should be able to adapt more readily to changing work structures and processes. About younger workers: "Microsoft relentlessly targets the brightest rising stars in the computer industry. The only companies that will thrive in the knowledge economy are those which recognize that people are the ultimate scarce resource."

contd.

203

Table 8.3 Continued

The world we live in	Human history, like the planet, runs on natural cycles. Future elders must be prepared to be more self-sufficient in retirement, but America should not undertake big benefits cuts. Generational teamwork will be necessary.	More elders than ever will mean the nation must confront intergenerational economic issues and protect the future. Some authors fear a crisis is coming; others insist: "The sky is not falling, the sky is not falling."	Peterson proposes major global summit on aging to find multilateral policy solutions to pension redesign, work structures, and international labor flow between developed and developing world.	Opportunities: "employment and investment strategies that could make a difference between affluence and poverty in old age; the big money to be made in ageing, particularly in the financial and leisure services and

Strauss, William, and Neil Howe. *The Fourth Turning: An American Prophecy (What the Cycles of History Tell Us About America's Next Rendezvous with Destiny)*. New York: Broadway Books, 1997.
Thau, Richard D., and Jay S. Heflin, eds. *Generations Apart: Xers Vs. Boomers Vs. the Elderly*. Amherst, NY: Prometheus Books, 1997.
Dychtwald, Ken. *AgePower: How the 21st Century Will Be Ruled by the New Old*. New York: Jeremy P. Tarcher/Putnam, 1999.
Wallace, Paul. *Agequake: Riding the Demographic Rollercoaster Shaking Business, Finance, and Our World*. London: Nicholas Brealey Publishing, 1999.
Peterson, Peter G. *Gray Dawn: How the Coming Age Wave Will Transform America—and the World*. New York: Times Books, 1999.
Roszak, Theodore. *America the Wise: The Longevity Revolution and the True Wealth of Nations*. Boston: Houghton Mifflin, 1998.
Walker, Jean Erickson. *The Age Advantage: Making the Most of Your Midlife Transition*. New York: Berkley Books, 2000.
Roszak, Theodore. *Longevity Revolution: As Boomers Become Elders*, Berkeley, Calif.: Berkeley Hills Books, 2001.

Endnote: Gray is Beautiful

Books about technology and work by no means dominate titles on aging. In combing bookstore and library shelves (physical and virtual) in search of volumes to consider, I found treatments of middle and old age as basic as *Retirement for Dummies* and *Sex After Sixty*. I also found a number of new-age celebrations of Baby Boomer maturity, whose authors mostly are women. Examples include *Red Hot Mamas: Coming into our Own at Fifty* and *Goddesses in Older Women: Becoming a Juicy Crone*. Countless "solutions" to aging are by far the most plentiful of the growing-older genre. Most focus on maintaining physical and emotional vitality and leading "successful" or "productive" elder years. Some mention capitalizing on the benefits of technology, either through lifelong learning endeavors or creating an electronic journal, for example. The "senior surfer" has become a prime audience for the Gray is Beautiful genre. *Fifty-plus and Looking for Love Online* is one example. Readers learn how to perform the work of presenting themselves as available mates in ways that are both safe and effective in the new electronic age. Finally, the old age "spoofs" are sprinkled into the lot. They focus on the humiliations of lost youth and quite possibly are most often purchased as birthday gifts for 50-year-olds (an industry unto itself these days). The elderly's perceived ignorance of new technology appears in virtually all of these texts, right along with the propensity toward flatulence. One example is a tiny-format volume by Emma Burgess called *You Know You're Getting Old When* . . . (2001). Each page is a separate joke, such as "You Know You're Getting Old When . . . You refuse to give into this PlayStation lark . . . until they bring out *Lara Croft Plays Hoop and Stick*."

CHAPTER 9
Driving with Dad: Intergenerational Journeys on the Superhighway

In Fall 2000, I was in my first year as chairperson and I had just received tenure in a small, contentious mass communication department at the University of Wisconsin–Milwaukee, an urban university with a large proportion of first-generation and returning college students. I celebrated my good fortune by giving up my sabbatical at the dean's request and agreeing to teach a section of the senior seminar in order to facilitate the departure of a colleague who was joining another unit. In addition, I was at the helm of a project to start a new campus center on age. My service responsibilities were taking me off campus as frequently as they allowed me to inhabit my office. My work life had begun to remind me of something I had once heard a visiting lecturer say in a campus workshop on teaching with technology: "I have to get through my teaching responsibilities so I can get back to my real job, which is 'Go to meetings.'" I surveyed possible texts to use and could not even get excited about requiring undergraduates to read *The Whole World is Watching* by Todd Gitlin, as I had done on previous occasions. Nothing rang the bell for me. *What better time to experiment with my teaching?*

My interests had moved on from television to new technologies and the Digital Divide, which presented a practical problem. Our department lacked an Ethernet connection in its cramped seminar room, so a dozen students and I elected to hold class in my office, which was equipped not only with Internet capability and an old monitor but also with a large table.

I developed a syllabus that I would never have dared to offer before being tenured. On the first day of class, the students were shocked to learn that there would be no printed readings, no books, no exams. *That,* they liked. Instead, they were told that they must identify the 55-plus-year-old adult "intergenerational partners" of their choosing and recruit them to participate in the sixteen-week project with them. The "partners" had to answer a brief questionnaire that sought demographic information, work history, and an accounting of their computing experiences. Yikes.

One student, about 40, walked out of the first class and dropped it immediately. She complained, perhaps justifiably, that her elderly mother had not signed on to go to college. The student was able to find a more traditional course with readings, term papers, and a final examination. That left me with eleven terrified students, late Generation Xers who were in their element with the Internet but whose 55-plus associates, they feared, would revolt at having to spend a semester with it. Going to the library to do research for a media-effects paper couldn't have been worse, they reasoned.

At that first meeting, I told the students that my current research was on the relationship between elders and new technology and that I was interested in their learning more about how older people use the Internet. Together, I suggested, we would learn more about this complex and exciting phenomenon. Despite their palpable relief about not having to read Todd Gitlin and the like, the students' disappointment shone through. The classroom was a collective snore. Old people!

Eyes glazed over as I walked through the Web sites mentioned on the syllabus and spoke of how we would use them via our hip, Blackboard-driven course. Each week the students would visit a cluster of Web sites with their partners and write about the experience. One week, the pairs would visit AARP's Web site and some of its links. Another, they would visit CNN's Web site and its Aging and Health pages. Seniornet and other sites catering to elder users were designated for another week. Financial Web sites, sites about technology news and information, and government sites about Social Security and aging were some of the other featured materials. Students entered sites with their partners through a direct link on our Blackboard site. At the end of the course, students would have to cap their weekly diaries with reflective papers that would analyze the Internet from a fresh perspective and make suggestions for how a future course might improve on this one. Weekly class time for the three-hour course would comprise class visits to the previous week's Web sites and discussion of issues common to the students' experiences. Class time also would include visits to Web sites that students found relevant to a discussion of intergenerational issues and experiences.

Expecting snowballing negative reaction from my early- to mid-20s students, I was unprepared when they started to drop by the office with a different message. Covertly at first, they delivered their messages as if they might feel embarrassed to "go public" with them. Several thanked me for the opportunity (to see my grandma once a week and show her what I do, to get my mom to "schedule" time with me). But the majority remained nonplussed, as far as I could read their reactions, until about four weeks into the semester, when the class conversation started to pick up steam. In addition to discussing research that my graduate students and I were doing with older users of technology and talking about the students' own, fairly homogeneous experience with computers, we discussed the weekly meetings with the partners, most of whom were parents or grandparents.

In about half the cases, these young people were first-generation college students. About half the elder partners mostly had never even touched a computer keyboard when the course began. Nervously, the students, with almost no guidance, stepped gingerly into their new roles of family expert on the future, holding an elder's hand while introducing them to the new century's communication forms. I held my breath.

The students began to trade "Can you top this?" stories about their experiences with their intergenerational partners. They related humorous tales with a sensitivity that began to bond them as a group. Vikisha told of spending hours with her grandmother before they could ever go onto the Internet, getting Grandma comfortable with a keyboard, screen, and mouse, for example.

> Things I never even consciously thought about, things I have grown up taking for granted, like using a scroll bar, seem awkward to her. I have to tell her over and over again about a simple step that will save her thousands of keystrokes. And sometimes I just don't mention it, because, well, I don't want her to start thinking I think I know so much. I want her to feel comfortable.

When such expressions came, as they did often, the classmates huddled in support. They began to notice that they were learning, not just about their elders' use of technology, but about their own. Such accommodations as the one Vikisha expressed were completely new to these Generation Xers, who had always expressed deference to their elders on some level, but certainly not technology. They were starting to understand the ambivalence—pride and anxiety—that Grandma felt when she noted that the kids had moved way ahead of her on any given technological front. They could still laugh about the elders' VCR flashing "12:00," but the image now held a poignancy for them.

The energy of the class progressed from nervous titters to joking, arm touching, and just plain noise. By Week 7 we were forced to keep the door closed, after a few colleagues and several wandering students had poked their heads in, wondering why class time was seemed overly social. (One unregistered student tried to join, but we couldn't allow it.) One of my colleagues expressed concern that there were no books, and, besides, we weren't in the therapy business. It had come to his attention that the class had grown to be about the students' changing perspectives on their relatives, and their relationships to them. Although I explained to this colleague that such events were byproducts of the course, not its main contents, he remained stern. *Jealous,* the students concluded.

The class period would not contain the gathering. The seminar participants continued to "hang"—ten minutes, twenty minutes, thirty minutes after class had ended. They started to form a community unto themselves, in a way that I had not previously experienced with a senior seminar. Everybody was graduating (if they all passed their law class). Everybody was looking for a job, and a few had one. Very often, they stuck around the office chatting about their "date" with Dad or Grandma, but very often not. Who was Vikisha's new boyfriend, and was he a student? Was Kate still sneaking cigarettes, even though she'd pledged during the second week of class to quit? Would Shannon pass the law and regs course so she could graduate? Was Ben very far from turning into a Republican, like his grandfather?

Three conditions distinguished the experience:

1. *The environment.* The class was held in my messy, informal office (with *great* purple chairs), the Internet clumsily hooked up on a monitor screen, and the phone occasionally ringing in the background. Students got up from the table and sat in my desk chair to call up Web sites they had visited with their partners or the Blackboard course site to refer to postings. The talk was informal and plain. I placed the success of the seminar entirely on them, and this created enormous peer pressure, which was unleashed frequently. The Milwaukee equivalent of a shaming would take place if the group got the sense that one of their members was letting them down through a lack of participation or preparation. A pride of performance kicked in, and a couple of students were disinclined to play. It was just like home, and that suited the course to its content.

2. *A call to lead.* Instead of contact with books and term papers, the required contact was with people they knew. The experience led them outside their comfort zone as students and as family members, where they were called upon to demonstrate their knowledge to an elder for perhaps the first time, to re-earn that elder's trust in a new way. They took on leadership roles and did so gently, elegantly.

3. *The relationships.* All seminars are organic, but the nature of the students' relationships with their intergenerational partners defined this one in a way that I had not anticipated. The relationships changed, some subtly, some more dynamically.

Each intergenerational team's experience was distinct, of course, but many points of overlap occurred. In their course diaries, the students held nothing back, choosing to maximize the course's function as a capstone "rite of passage" into a state of adulthood they had not previously known. Six summaries among the eleven highlight the class's encounters with technology and their teammates.

Student: Adam, 22, from small-town northern Wisconsin, white.

Teammate: Dad, 55.

Previous computer experience: Basic computing experience, no Internet experience.

Relationship: Dad was noncustodial divorced parent, now living with girlfriend, her 15-year-old son, and Dad and girlfriend's 11-year-old daughter.

Partner meetings: Weekends, phone meetings.

Starting the project: Adam feared being a "toadie" in a research project and resented the course material—"old people."

Partner choice: Wanted to spend more time alone with Dad.

Obstacle: Reaching Dad. Adam was away at college and had to rely on the phone often for their meetings.

Adam's challenge to Dad: "Do your banking online. Use the Internet for commerce. It's safe." Second, Adam persuaded Dad to get his home computer connected to the Internet for the benefit of his younger children. This "overjoyed" Dad, who had been feeling anxiety about the kids' Web-related education at school.

Dad's challenge to Adam: Having mastered basic Internet skills, Dad turned to these to demonstrate the need for Adam to begin an investment program. Adam got an online bank account and began purchasing stocks after Dad and he learned how to do so using the Internet.

Biggest surprise: Dad showed an interest in information that he and Adam found relating to prostate cancer. This was the first time Adam perceived his father as vulnerable due to aging.

Conclusion: "My Dad was afraid. This is exactly how I felt when the Internet craze first hit, and they are the same feelings I have now when I see online investing and business ads. It is the feeling that you are being left out of something that is too hard to learn and therefore not worth your time, and it is not part of your life. When you finally do invest time to learn about a new technology, you begin to realize the potential of it and soon you forget how you lived without it. That is exactly how my Dad felt at the end of our semester together. He would use the Internet confidently at work and at home now."

Student: Melissa, 25, southside Milwaukee native, white.

Teammate: Grandmother, 73.

Previous computer experience: Never used a computer.

Relationship: Distant but friendly. Grandma was Dad's mother.

Partner meetings: Weekends, with a special Sunday dinner at Melissa's parents' house as the "carrot" for Grandma.

Starting the project: Fear and anxiety for both partners, Grandma about computers, Melissa about being the "boss."

Partner choice: Melissa wanted to spend more time with her grandmother, who was a "strict family matriach." She feared imposing on Grandma's time but was delighted when she agreed.

Obstacle: Grandma preferred the Packer games on TV. Also, Melissa was nervous about being the teacher to Grandma. She feared that something would go wrong and frustrate Grandma, upsetting her at Melissa. This did not happen.

Melissa's challenge to Grandma: Get comfortable. Grandma had to work to learn to use the mouse and scroll bar, recognize a link and how to reach it, know she must wait when she sees the hourglass.

Grandma's challenge to Melissa: "Now that I've learned the Internet from you, I have some things I want you to look up for me." (No interest in revisiting now that her homework is done.)

Biggest surprise: Grandma paid close attention to the exercise graphics she found on the National Institute on Aging Web site. She also showed interest in yoga and Tae Bo. Also, Grandma showed keen interest in Web sites related to seniors' chronic illnesses. Grandma had always seemed healthy and vigorous to Melissa and it stunned her to see this new, vulnerable side.

Conclusion: "Grandma will probably go the rest of her life without using the Internet again. But she learned a lot more than she realizes. She is more computer savvy in general now, and she is aware that there is much more there than pornography. Seeing a Web site listed on the side of a product package will no longer mean a mystery to her."

Student: Alisa, 33, central city Milwaukee native, African American.

Teammate: Mother, 57, single mother of three, deputy clerk of Milwaukee County Court, some college experience.

Previous computer experience: Computer literate but no Internet experience.

Relationship: Very close family.

Partner meetings: Whenever Mom could find spare time.

Starting the project: Excitement. "Mom is a lifelong learner."

Partner choice: As the eldest of the three children, with the youngest just 18, Alisa was ecstatic over having an excuse to have her mother all to herself.

Obstacle: Alisa's mother had little time for the project, but Alisa persisted, and they spent a few hours each week on the computer.

Alisa's challenge to Mom: Get comfortable with the Internet. Mom took a while to learn how to use the Back key, for example, but showed lots of interest in using the White Pages to locate a relative in Kansas City and for related queries.

Mom's challenge to Alisa: "Get me away from these senior sites!" However, Mom became attracted to medical sites, where she could learn more about her condition, which had been newly diagnosed—spinalsclerosis.

Biggest surprise: Mom showed up at Alisa's apartment for the first Intergenerational Partner meeting with pen and notebook in hand. She was going to school.

Conclusion: "I have probably given my Mom too much enthusiasm. She will show up at my home any time of the day or night now to get on the Internet to look for information she wants."

Student: Brandon, 25, small-town central Wisconsin native, white.

Teammate: Stepfather, 56, computer science teacher at Wisconsin Dells High School, just before early retirement was to occur.

Previous computer experience: In addition to twenty years of teaching high school students basic computer skills and applications, developed his school district's first Web site several years earlier.

Relationship: Brandon seeking stepdad's approval.

Partner meetings: Weekends and on the phone.

Starting the project: Brandon was nervous because of his stepdad's pre-existing knowledge about computers. He knew this assignment focused on the Internet, though, and that was an area Dad had not used much.

Partner choice: Brandon knew no one else in the required age group well enough to ask for help.

Obstacle: Dad balked at the early required Web site visits focusing on aging and retirement, concentrating instead on the design of the sites. Brandon had a revelation: "I'm beginning to see a pattern in the way he's looking at these Web sites: He's avoiding the subject of aging. He's in denial about his impending retirement. He's amazingly unprepared. He's feeling like he will become one of the old Tandys the school district first bought. He's scared he will lose relevancy when he retires, because he will be unable to upgrade and become obsolete and unwanted. Losing his teaching career will be unsettling."

Brandon's challenge to Dad: Use the AARP site to prepare yourself for retirement. As of semester's end, Dad resisted. He did not want to be associated with AARP, even though he took advantage of its member discounts.

Dad's challenge to Brandon: Stray from the assigned Web sites—Boring!—and look at some of the Christian bookmarks I have on the computer. Brandon did not enthuse about the subject matter but approved of his stepdad's creativity and resourcefulness with the Internet in seeking out and organizing this battery of sites.

Biggest surprise: Over the course of the semester, Dad reversed his decision to take early retirement. In part, the experience informed him that he was not ready to do so.

Conclusion: "Dad needs his projects. Even when he does retire, the Web sites will be a big part of his life. I don't think he will go cold into retirement but step down toward it gradually. He will need to make a series of adjustments to maintain his identity. The course experience was great for us, because I saw more of my parents. Parents are happy to help you with this because they just want to see you at this point and want to know what's going on in your life. Personally, I was shocked when the student walked out on the class. She shouldn't have worried about someone rejecting the proposal of helping with the project. Rejection is part of life. You go on and find someone else, or you explore why they are reluctant to look at the Internet. That's just as important as finding out why people use it."

Student: Shannon, 23, upper-middle-class suburban resident, white.

Teammate: Father, 57, physician at medical college teaching hospital.

Previous computer experience: Fairly experienced Web user. Had bookmarked about a dozen sites, including the Weather Channel, America's Cup, Mapquest, Jerry Baker's Gardening, and Microsoft Micromedic, a specialty site geared toward surgeons.

Relationship: Close family, but Dad, a busy medical professional, had little time for Shannon after work and golf.

Partner meetings: Whenever Dad could spare up to an hour, which was not every week.

Starting the project: Defensive Dad balked at the AARP and related sites. Shannon found this amusing, given that he had started to depend on bifocals to see the computer screen, lavished praise onto pages that were easy to see and read, and became excited by the travel deals he found on the AARP site. Dad began to apply some of the available information there—Alzheimer's features, for example—to others in the family, his mother and hers, resisting any association with himself.

Partner choice: Shannon reasoned that, since her parents lived 20 minutes away, she could do laundry and homework simultaneously and still get dinner.

Obstacle: Shannon felt forced to "mingle" with students who were not like her in the seminar. The setting was much more intimate than any of her other classes had been. "We all share something now. We worked together all term, and we compared stories about working with our partners. It seemed at the outset that the experiences were really different, like with some of the partners being elderly and others being our parents' age, older Baby Boomers. We learned things we had in common, and we learned respect for each other. Young, old, black, white, male, female—I walk away from this with a firmer grasp of the world around me. This is a great experience to end college with."

Shannon's challenge to Dad: Examine Web sites you didn't know about, like the U.S. Administration on Aging. Look closely at the National Institute of Aging Web site. Learn about the resources that are there. Dad ultimately found that the experience enhanced his medical practice through new resources for himself and his patients.

Dad's challenge to Shannon: Learn that some health information on the Internet is good for patients, but that data and advice about complicated conditions can do a disservice. Medical sites should not interfere with health care professionals' roles in taking care of patients.

Biggest surprise: It turns out that Dad receives *Modern Maturity* magazine from AARP and reads it along with a variety of other publications. Shannon learned this from Mom.

Conclusion: "The class brought us closer because we talked about a lot of things. I came to understand that he was indeed getting older. He does not want to be. I think that is why he was so hard on the AARP sites. He did enjoy showing off his medical knowledge to me. And our house has always been a gadget one. We had a Commodore 64 circa 1983. Now my Dad has moved on to a DVD player. I like seeing how he fits technology into his work life just the same way."

Student: Guy, 24, southside working-class Milwaukee neighborhood native, white.

Teammates: Nana and Papa, in their early 80s. Guy's grandparents had emigrated from Italy in their youth. Papa had completed five years of formal education and Nana even fewer. She had gone on to raise six children in their conservative Catholic, working-class family.

Previous computer experience: They had never touched a computer and had no desire to do so.

Relationship: Large, close-knit family.

Partner meetings: Weekly, in grandparents' home. Guy would haul over his sister's computer in boxes before the latter part of the term, when someone loaned him a laptop.

Starting the project: Complete dread for the grandparents. For Guy, guilt, compounded by his sister's and mother's criticism of him for "taking advantage of your grandparents to help with your homework when they are so frail and elderly. Why did you not pay such attention to them before?" I would like to say that these conditions changed, but they only grew worse. The only highlight of the experience was when, late in the semester, Guy brought over the borrowed laptop, with its superior Internet connection, and smoothly transported them to Web sites showing pictures of their native Italy. They were temporarily fascinated but then became frustrated over not understanding how the laptop worked.

Partner choice: Guy asked his grandparents to partner with him but immediately wished he had picked his mother instead.

> **Obstacle:** Lots of interference from the extended family. Plus, Guy compared his grandparents to an older version of George Costanza's parents in the *Seinfeld* show. They would argue, for example, over who could see better and who had better hand-eye coordination for working the mouse; in other words, who got to drive. Papa won but Nana heckled him constantly. They often became frustrated by the computer's refusal to cooperate with their skill level. Ultimately Guy "drove." He volunteered a chart measuring his grandparents' engagement of the technology in five partner meetings.
>
> **Guy's challenge to Nana and Papa:** Touch the mouse. Make it go.
>
> **Nana's challenge to Guy:** "Take your muddy shoes off and don't drag that thing in this house again. It has nothing to do with us."
>
> **Biggest surprise:** "The Internet is a fraud. I thought it was truly 'inter'-active. But it doesn't welcome everybody. It just made my grandparents feel more excluded. They're doing just fine without the Internet. It doesn't carry the handwriting of a letter, the inflection and volume of a voice, the things they examine for cues in their communication with others. I think a lot less of the Internet since I tried it out on my grandparents."
>
> **Conclusion:** "The differences between my grandparents and my peers are just mind boggling. The stereotype would be that they are uptight and all rule abiding. I believe they are actually more laid back. We get all stressed out by the pressure of keeping up, and a lot of this involves our interaction with computers. I am not smarter than my grandfather. I can't imagine learning more than he knows now."

On the last day of class, Ben gave me my first dreidel. Vikisha announced that her 80-something grandmother, "my best friend," had surprised her by enrolling in a computer literacy class at her local senior center:

> Honestly, I hadn't expected that she would ever get on the Internet again. She had started to make lists of things for me to look up, like sites about genealogy and unclaimed money. She has come from feeling like, if she touched the computer, she would damage it, to wanting to use it on her own. I think she is looking for something on the health sites that will tell her it's OK to eat chocolate and nuts even though her cholesterol level is high. Also, she is starting to locate information about women and smoking. No doubt, she is doing that for me.

Ben, a 23-year-old reform Jew from the affluent suburbs of Milwaukee's north shore, had had the least impact on his partner's computing practices, he guessed:

> My grandfather is a 78-year-old retired general practitioner. He uses the Internet every day to keep track of his investments. I was intimidated about showing him.

But then I realized that he did not know some of the basics, like how to use the Back key. I politely steered him to learn these things.

Ben received approval from his classmates when he put the intergenerational contact in the context of his age cohort entering the work and social world:

> The only real way to understand a technology is to understand the whole culture. This means young and old alike and should include everyone in both groups. What has become horrifically apparent is that the opinions and feelings of older generations have been discounted when it comes to the Internet—since its conception. That is not to say that things haven't moved away from this absolutist stance. With Web sites like AARP, many people will become less disenfranchised in the realm of computer technology.

The students agreed that they all would go out to the workplace with an improved appreciation for intergenerational communication, differences, and common ground. They felt these assets would help them get along better in a society that required escalating capabilities with group use of technology. Several also took solace in the realization that their still-working parents were better positioned to work in this arena, too.

As Kate noted, "While computers are amazing to my mother's [leading-edge Baby Boomer] generation, they're also scary as hell. She didn't even have a TV set until she was 11 years old. But she is a lifetime learner. She is 55 and delighted about the opportunities she has. What a role model."

For Kate and her classmates, the grandparents in the group largely emerged as "elders," a group that did not have to adjust to the Internet life but could do so in pursuit of self-fulfillment. The parents presented a more complicated picture: Vulnerable workers with enormous potential, perhaps an element of denial about their own aging, but with a quiet grace. "They're smart," Erik said. "They're still open to learning. Many of them know they need to learn."

I left the course feeling optimistic that these eleven young people were destined to be good intergenerational teammates in their workplaces. When I try this approach again soon, I will make some improvements, most of which were suggested by the students in their final papers. For one, the course will begin and end with the older adults in the classroom with the students. This exercise seemed impractical at the time, but, to maximize the communal function of the group, it seems necessary. For another, I will invite more freedom among the surfing teams. Many of them simply hated the "aging" sites, not only because of the discomfort they caused but because they found these "boring." One week of "aging" sites will be sufficient, and then I will encourage the teams to explore sites more suited to their own tastes.

A footnote: Toward the end of the course, I received a strange email in the office on the day of my seminar meeting. It came from someone called IrBigJoe. It read: "Hey, Karen, It's Grandma. Nothing's wrong. I'm over at Uncle Joe's and he is showing me the Internet. I love you. Everything's all right. Grandma. I love you."

When my students saw the email, from my 91-year-old disabled grandmother, they cheered.

Mentoring Relationships: New Directions

In the past two decades, a proliferation of scholarly research and popular business communication has grown around the theme of mentoring. The majority of the literature has suggested, in essence, that inexperienced people (read: young) should find experienced mentors (read: mature) in order to ensure patterns for success in the workplace and in other aspects of life.[1] Pamela Kalbfleisch and Arlyn Anderson note that the first "Mentor" was the mythological surrogate parent of Odysseus, who taught the young adventurer how to find his way in the world. Kalbfleisch and Anderson's research demonstrated that grandparents commonly serve as mentors to their grandchildren, who function in protégé roles. This allows the elders to enjoy continued interaction in a helping function as they demonstrate and instruct the younger family members in work skills and education matters.[2]

Traditionally, this model has served societies well. As a matter of convenience, such demonstration and instruction have been handy for older adults to extend to children who have often been in their care and who more often have been living nearby. However, as the pace of life escalates for activity-filled teens and young adults, and as they assume greater familiarity and self-confidence with newly valued skills and knowledge, the strictly top-down approach to cross-generational mentoring is rendered unsound.

Instead, a dialogic mode of mentoring is needed now, more than ever before. Elders retain a place of significance in the transmission of culture. They have lived through decades of experience and change and in all cases have much about which to instruct their young. On the other hand, the younger Generation Xers and Net Generation behind them have much relevant skills and knowledge to impart to their less experienced elders. Richard Wiscott and Karen Kopera-Frye have concluded from their research that grandchildren and grandparents often instrumentally work to help one another understand the dynamic elements of their shared culture. While grandparents are most often involved in transmitting heritage to the grandchildren, as Wiscott and Kopera-Frye note, I am emphasizing that the younger partner's sharing of up-to-date cultural aspects can also help the two achieve a shared perspective.[3]

My students at first felt trepidation about assuming the mentoring role with their elders. After all, these were the people who had changed their diapers, taught them to not run in traffic, and, perhaps above all, taught them to respect their elders.[4] But, to a person, they each soon relished the new role, for the opportunity it offered them to "have an excuse" to be close to the dear relative and to refund in some way a debt they felt they owed.

Reciprocation has traditionally been a primary motivator for adult children's and grandchildren's contact with their elders. That reciprocation has often taken the form of care giving or other provisory role that recognizes the elder as in decline and in need.[5] Younger people are poised now to reciprocate in a way that will allow their elders access to a world that many feel is increasingly leaving them behind. This is especially true for those elders who are past the age of paid work, where they are likely to encounter new technologies as part of the structure of their lives rather than as a present absence in their lives, a show to which it seems everyone has been invited except for them. As people enter old age, they are likely to embrace methods of retaining their competence and autonomy. Now, as Melissa noted in class, when her grandmother is in a conversation during which the Internet is referenced, she can at least speak with some authority about having "been there." Melissa added, "I truly don't expect Grandma ever to go back on the Internet, but the experience has made her better off because now she isn't at a loss about what it is, what it looks like, what you do with it."

Intergenerational Empathy at Work

Several of the students in my seminar noted that they felt it would be easier for them to go into a mixed-generation workplace and work effectively because of the experience they had sharing technology with their parents or grandparents. "I saw that what is second nature for us has to be real conscious for them," one of the students said. Poor ergonomics in the desktop setting, poor lighting, bug-ridden systems, and problematic interfaces are things the Net kids take in stride, but they can make the difference between an elder becoming a competent user and giving up, the students agreed.

These students were able to observe firsthand results of what Sara J. Czaja and Chin Chin Lee have identified as the frustrated encounters some older adults make with new technology. Although they are receptive, they often encounter barriers to successful adaptation, and these barriers can be sufficiently discouraging as to put the elders off the technology for good:

> Overcoming these barriers depends on training and design solutions that accommodate age-related declines in perceptual, cognitive, and motor abilities. This might involve, for example, software modifications, alternative input devices, or redesign of instructional materials. The development of these solutions requires an understanding of the needs, preferences, and abilities of older people.[6]

When Kate or Adam or Vikisha becomes a unit manager or Web developer, I will venture that intergenerational computing concerns will be a more prominent concern than might otherwise have been anticipated.

Intergenerational partnering experiences such as the one in the Milwaukee seminar certainly will carry rewards, but I am interested in seeing colleges and universities take up the issue of intergenerational studies more definitively, both through general education and across the curriculum. The demographers have taught us nothing in the past several years if not that we need to live more purposefully in an age-integrated society. College life, while more integrated across age groups than ever before, remains remarkably insulated as an experience for young adults. At my own residential campus in southeastern Ohio, for example, one would be hard pressed to find an undergraduate over 24 years of age. As a result of this cocooning and perhaps in spite of all the good diversity perspectives we as faculty toss out at them, students can graduate from college with narrow perspectives about the role they will play with technology in the workplace. Students in the intergenerational-partnerships seminar could see, however, that using the Internet, at least, is a set of practices that they formerly took for granted but that are, in fact, embedded in complex social and physiological circumstances.

I am asserting that if we as faculty members become more creative in our implementation of diversity initiatives for students, accenting intersections of technology with age, disability, and cross-cultural difference, we can help build a runway for a more cooperative model of work that soon will be led by today's students. This is not to say that reading *Beloved* in Junior English is not a democratizing experience, only that the diversity perspectives we instill through the humanities and social sciences should be galvanized with practical and interactive work in such disciplines as business, social work, and telecommunications. More, not fewer, such experiences are needed to bring into focus the ever-more-complicating picture of work in today's developed world.

Matters of intergenerational equity in the office will grow more urgent as workers stay at work for more years and workplaces become increasingly age integrated. Competition for resources in destabilized economies are bound to spark new fronts of conflict.[7] With new technology's overwhelming emphasis on processing speed at every turn, it will be tempting to scorn at the aging bodies that are judged to be the embodiment of speed's opposite.

Scholars of critical gerontology have provided overwhelming evidence of the marginalization of the aging body so that its appearance in the fast lane of the techno-hip office can be both surprising and unsettling. Guerrilla rhetoric from some of these same scholars proposes resistance of externally imposed meanings of aging bodies (declining, too slow) through a process of reclamation of control that is now held by younger people.[8] As Emmanuelle

Tulle-Winton suggests, "In cultural terms, this may involve transgressing aesthetic boundaries and accepting, rather than shunning or marginalizing, the biological signs of old age."[9]

The classroom and its prostheses—the service learning community, the distance learning course, for instance—can be important sites on which to wage this transgression. College students who learn to face the realities of aging bodies in their encounters with technology will be invited to achieve a phenomenological understanding of what it means to grow old in contemporary society. As these same students go on to interact with elders in many settings of their working lives, this head start will be of benefit in countless ways.

Technology Skills Acquisition for Enhanced Aging

Social scientists have demonstrated that older adults who use computers to sharpen and maintain cognitive and sensory skills more likely will enjoy an ideal learning context because they will have performed activities they find useful.[10] Researchers also have accumulated ample evidence of older adults' desire to learn to use new technology, including the World Wide Web, and that their chief reason for not using the technology was their lack of knowledge about how to do so.[11] Wendy Rogers and her colleagues advocate a system-based training approach for older-adult computing trainees rather than a program-driven approach. This means starting with an analysis of the user and his or her computing needs, then tailoring a program around those specific circumstances.[12] Such an approach is oppositional to the routine of computer training, which is standardized.

The intergenerational-partnerships approach, while it was not a formal computer training experience, did involve tailored settings. The young adult children or grandchildren began with an informal assessment of what their partners knew or wanted to know and proceeded outwardly into the elder partners' fields of interest.

An approach that centers on the older learner, rather than on the training design (just add water!), seems especially crucial in the increasingly diverse scenarios in which these adults are coming to computing and the Internet. Physical and cognitive impairments are common with those over 60, but these specific impairments and their rates of incidence vary. People with age-related sensory impairments and cognitive losses sometimes can be helped by appropriate exposure to computing technologies, both in terms of recovery of some abilities and through recovery of a lost sense of being in the social world.[13] Cookie-cutter training designed for "the older learner" tends to meld the issues of bifocal-wearing 40-somethings who might have years of computing experience with those of memory-losing 70-somethings who might just be coming to the technology. The multifarious nature of the experience of aging will become increasingly salient as computing enters the everyday lives of so many new kinds of users.

The "intergenerational partnerships" seminar helped to highlight applications of young-to-old mentoring partnerships, intergenerational-diversity perspectives as part of the higher-education experience, and the need to address computer training for older adults in diverse ways. As Ronald Mannheimer has noted, intergenerational-education projects not only help young and old overcome their own prejudices about the other but construct bonds between generations as they find common life themes and challenges.[14] In Chapter 10, I will return to the themes of intergenerational computing and education within the context of social policy.

CHAPTER 10
The Digital Divide's Gray Fault Line: Aging Workers, Technology, and Policy

> Every month, about a million people in the world reach the age of 60, and some 80 percent of these live in developing countries. In rural areas of Africa, Asia, and Latin America, the number of older people is supposed to double by 2025.... It took France 140 years for its older population to increase from 9 percent to 18 percent. It will take Venezuela just 22 years for a doubling of its aged population....
>
> Through a combination of good health habits—eating right, exercising, not smoking, and good genes, many people are not just living longer but living healthier lives. So how can we take advantage of their skills and experiences as employers, employees, as owners of small businesses, as active volunteers performing community service?
>
> —Esther Canja, President,
> *American Association of Retired Persons*[1]

As noted in the *Falling Through the Net* series, a study of Internet access among U.S. citizens sponsored by the National Telecommunications and Information Administration, the workplace is the location where most Americans over 29 first became computer and Internet users.[2] Most programs that address closure of the so-called Digital Divide focus on young people as targets of education, and that often means the K–12 system. This book has drawn another pathway to understanding the Digital Divide, which has been characterized by scholars largely with respect to particular groups' have or have-not status along the lines of access to technology, technological literacy, and capital to use the technology in meaningful ways. My interest in the Digital Divide is on the places where agency is being negotiated for people who are shaping the new processes of work and being shaped by the new modes of work of an electronic society. I am interested in what is happening in the lives of people who are older than the young techno-class of Generation X-men and -women who feel at home in these new work spaces, because they were raised according to the new models of change, hyperspeed, and decenteredness. The older generations who totter around the segment of the Divide I am watching are getting used to the new models but may not feel "at home" in them. In a way, because of their need to rev up, adjust, and participate in the invention of new work logics, these elders are much more adventuresome pioneers of electronic culture than the younger workers, who never learned to operate in the old, stable environments and did not have to adjust to their disintegration.

In this chapter, I will address the policy-making arena as it can stimulate and reward those older workers, whose roles in the "new work" are both vital and threatened. My research has concentrated almost entirely on the United States, and the suggestions I have are primarily local in that regard. But I hope that scholars can work globally to attune policy-making efforts to the need for inclusive models and fair practices that will result in meaningful intergenerational experiences.

Despite politicians' euphoria to the contrary, the scholarly community sees no end to the Digital Divide, whether it is cast as a sheer matter of access to Internet hookups or as a more culturally complex issue involving a population segment's interest in, use of, and proclivity toward adoption of new technologies. Among the factors most actively complicating the Divide is age. On a simplistic level, we know that, at least in more developed countries, people between 50 and about 70 are less likely than younger people to use the Internet and to own computers, but they are doing so in increasing proportions. People in their late 70s and beyond are unlikely to become Internet adopters, although some do, especially if they are members of the upper middle class. In the United States, class and race or ethnicity help to predict on which side of the Digital Divide elders are positioned, along with income and educational levels.[3]

Similar factors distinguish the experiences of elders in more developed and less developed countries. Elders in North America and Western Europe, for instance, are much more likely to be Internet users than elders in other regions. (Of course, they are also much more likely to have access to telephones, electricity, and other conveniences of contemporary life.) Even within the more developed world, however, distinctions exist. Pensioners in Europe, who must pay by the minute for telephone modem use, are more likely to avoid the Internet than their American counterparts, who enjoy near-free or flat-fee access. But within Europe, pensioners living in Scandinavia are more likely to use the Internet than elders of most other countries.[4]

In the United States, where home computer and Internet usage are present in more than half the nation's households, work status directly influences usage, especially if you are over 50. Labor force participation significantly influences computer ownership and Internet use among Americans aged 55 and older.[5] For example, if you were in the U.S. labor force in the year 2000 and you were 50 or older, you were three times as likely to use the Internet as your non-wage-earning cohort members.[6] Even though computer-user rates are growing fastest among Americans 50 and over, this age group remains the least likely to use computers and the Internet (discounting 3–8-year-olds).[7] Currently, rates of computer and Internet use climb steadily from childhood until the prime workforce years (20s through 50s), level off, and then fall after about age 55.[8]

This plateau effect among elders will disappear as currently computing Baby Boomers continue to advance in age (cohort effect) and more new users continue to adopt the technologies at all ages.[9] Adoption clearly will come more slowly to Americans who are currently retired, have small incomes, and have not previously been exposed to computers or the Internet through the labor force or other means. This means that, on the whole, African American, Hispanic, and some other nonwhite elders will experience especially disproportionate adoption rates.[10]

Effective public policy for curing the Digital Divide must include attention to these Americans, many of whom are single women, racial minorities, and residents of central-city or rural geographic areas.[11] One means by which the Divide can be addressed is through policies that encourage the participation by elders in the workforce to the degree that they desire and need to do so. Training programs that help older workers remain current with technology and help nonworking elders retool to fulfill desires of re-entering the workforce can provide a much-needed missing link between the technology haves and have-nots in their 40s, 50s, 60s, and beyond. Several of the central-city Milwaukee women whose stories are told in Chapter 7 underscore this claim.

The aging of today's Baby Boomers will do little to address the Digital Divide in more developing nations. Older adults living in the developing world are especially likely to belong to the Digital Divide's "have not" group. In a study of elders living in the Asia-Pacific region, researcher Anne Cooper-Chen found that people 65 and over are much less likely than younger people to use computers and the Internet. She found four general reasons: the likelihood that they neither knew English (the dominant language of the

Internet) nor were familiar with Western language characters; the probability that they were not well educated and therefore discouraged from learning new technologies; the likelihood that they lived with their children, obviating the need for the Internet's perceived social benefits; and their more easy adaptation of cell phone technologies, which represented a more familiar communication form.[12]

As Katrin Markus of the European Federation of the Elderly has noted:

> The digital divide means that you have on one side people who are informed, and you have on the other side people who are not informed about what has happened in the world and what has happened in communication and information.... And to bring both of these columns together, the policy has to be strengthened and the industry must... engage themselves.... Forty percent of the EU population over the age of 50 have ever used a PC.... And from these 40 percent, 20 percent are over 70, and 10 percent are over 80.[13]

In studying an aging world population, the developing world is where the action is. We tend to think of aging societies as a Western phenomenon (as well as Japan), because the Baby Boom is perceived in such terms. The world's "oldest" nations today are, in fact, in the developed world; in such societies, within decades, we may see families that have more elders than children. But the upward creep of age is nearly global, with more than half the world's 65+ population now living in developing nations.[14] Persistent lowering fertility rates, increasing longevity, and, in the case of many developing nations, emigration of younger citizens to more developed countries, are enhancing this trend.[15] Kevin Kinsella and Victoria A. Velkoff, in their U.S. Census Bureau analysis of the aging world, highlight global aging as movement toward an understanding of the planet as a set of societies with interlinked generations, needs, and solutions. They note:

> We are all part of an increasingly interdependent and aging world. Current growth of elderly populations is steady in some countries and explosive in others. As the World War II baby-boom cohorts, common to many countries, begin to reach their elder years after 2010, there will be a significant jump by 2030 in the proportion of the world's population that is elderly.[16]

Policies that grapple with the technological needs of an aging world will need to be international in scope but locally meaningful. They will require increasing attention from national governments and international nongovernmental organizations—such as the United Nations and International Labour Organisation (ILO)—as policy regulators, enforcer watchdogs, and propagandists (in the best sense of the word). Existing laws and policies vary but generally do little to promote technological literacy and access for elders as citizens and workers. Improved policies that promote adoption of technologies by elders must be sufficiently nuanced to acknowledge cultural differences in experience along such axes as disability, gender, and social class.

Disability is a crucial issue. As others have pointed out, new technologies can serve as an important equalizer for disabled citizens and workers—but they can also serve as additional barriers.[17] Thirty percent of Americans between 50 and 64 and more than half of those 65 and over have one or more disabilities (difficulty walking, difficulty seeing, difficulty hearing, difficulty grasping, or learning disability).[18] Among all 50–64-year-olds with a disability, more than half have never used a personal computer, compared with less than 30 percent of those in the same age group with no disability. The gap for those 65 and over is also severe.[19]

Increased levels of training for technological literacy among older adults in the workplace will not only increase successes for them as workers but will furnish enhancements later if they find themselves retired with a disability. As argued by the advocacy group Age Concern England, a policy group focused on bettering the lives of aging workers in the United Kingdom, skills with new technologies learned now can be an investment for the day when elders face unknown circumstances:

> The ability to communicate easily, cheaply and fast holds out hopes for decreasing feelings of isolation while the internet provides opportunities for overcoming problems of mobility, unavailability of transport facilities and can reduce the need to make journeys for shopping as well as providing a wide range of leisure activities.[20]

Gender critically informs the economic status of older women in the contemporary aging world. Women in their 50s and beyond, who are much less likely than men their age to participate in the official labor force and who are much more likely than either men or younger women to live below the poverty line, are unlikely to see improvement in their diminished health, workplace status, and financial resources.[21] Significantly, older women around the world do participate in considerable numbers in the unofficial labor force, however. While they are much less likely than younger women to work in the Information Technology (IT) sector or work with new technologies of any kind, they likely perform routine duties related to the family that undergird the computer-related productivity of their society. In more developed countries, women's work force participation at ages 55–64 is sharply increasing. For example, only about a third of Swedish women in this age group were in the labor force in the early 1970s. By the early 1990s, well over half were in the work force.[22] In many contexts, labor force participation by older women is especially necessary, with a widening gender gap in old age caused by a greater increase of longevity for women. This is complicated by the increasing prevalence of single older women living alone in the developed world, as divorced and never-married Baby Boomers begin to populate older groups.[23]

The need to work in the so-called retirement years is much more pronounced among African American women and men than it is for their white counterparts, whose net worth tends to be much greater.[24] Increasing sectors of older Asian and Hispanic American workers are expected by 2008, reflecting migration and birth trends. Such changes will contribute to an overall aging workforce; because recent birth rates have been quite high in Hispanic and black sectors of the population, however, the overall picture remains one of an older, white workforce and a younger one of color, with women figuring increasingly prominently in both groups.[25]

For men, significant differences exist. First, in the developed world, a two-decade-long trend toward "early retirement" has now stopped or, in many countries, even reversed.[26] In the United States, for example, 73 percent of men aged 65 and over were in the labor force in the early 1970s. That number fell to 55 percent by the early 1990s but now is slowly climbing again, as men increasingly find retirement systems insufficient to support them, especially in light of increasingly expensive lifestyles favored by Baby Boomers and a less romantic stereotype of early retirement as a golden time.[27] In fact, many men leave one occupation in their 50s or 60s only to take on new forms of work that allow them flexibility of mobility and income. For example, the pages of Web sites devoted to professional consulting services are clogged with the résumés of middle-class white men who have left corporate positions (or been downsized, as reflected in Chapters 2 and 5) only to seek new opportunities as their "own boss" or through short-term contracts with organizations. An increasing trend is for such men to retire from a company and continue in service to that organization on a contract basis, saving the corporation money on benefits that the retiree no longer needs and unshackling the worker from the requirement of being on site full time.

Policy groups such as the International Longevity Center–USA and the United Nations argue for official remedies that will pave the way for greater official labor force participation by older persons, especially women and those with disabilities, because such government and corporate protections will decrease the likelihood that elders will live in poverty and be denied access to the so-called good life.[28] Among the remedies mentioned are such initiatives as preretirement skills training for middle-aged women, who might stay in the labor force if they are better prepared for changing employment opportunities, and a transparent benefits structure, which will prevent some people from retiring before they can actually afford to do so.[29]

Such remedies will need to be introduced within policies that are designed to counter widening educational disparities between the generations, as the Rand organization, a public policy group concerned with social equality and fiscal fairness, has argued.[30] Even though the current generations of elders are better educated than any of their predecessors, they are, in many cases, competing in the workforce with junior employees who are themselves increasingly better educated. As a related matter, elders potentially find themselves on the wrong side of the Digital Divide, across the chasm from younger generations who came of age in a computer-literate society.[31] If significant changes in approaching the continuing education of older workers in the area of information and computer literacy do not occur soon, the "graying" of the developed world, especially in countries such as the United States, will further underscore the ravages of the Digital Divide. Even as more people 50 and over take to the Internet, digital inequities among the generations are made more profound, as concentrations of resources are pumped into K–12 education for young people while older adults receive scant "training" opportunities.[32] Furthermore, as the proportion of ethnic minorities grows among an aging population in such countries as the United States, digital disparities among aged citizens will increase: People who will be 50 and over a decade from now are, on the whole, more likely to be well educated and technologically literate than the previous generations of elders, but this is much less true for ethnic minorities who remain disproportionately represented in the underclass.[33]

Policy Solutions

Questions about the need for redesign of Social Security funding schemes are beyond the scope of the work here. But I will say that over-dependency on personal saving and annuity plans unfairly perpetuates the tendency for society to "blame the victim" for dependencies and hardships that may result. Obviously, people with better education and higher income are much less likely to be hurt by such a scheme than others.[34]

That said, I suggest that reasons do exist to lengthen the life of paid work, at least for many people. One of those reasons is the strength of the dependency argument made by reformists, who effectively note that a larger-than-ever schedule of retirees, paired with a shrinking work force, cannot be sustained by current retirement schemes such as Social Security. As these reformists argue, Social Security and other retirement plans, when conceived, did not anticipate explosive growth of the "oldest old" citizens. Workers who retired at age 60 or 65 could expect to live ten to fifteen more years if they were fortunate. My own grandmother has been collecting benefits for more than thirty years.

Social Security and disability benefits should remain available to those who need them, but healthy workers who can contribute a few additional years in the paid work after age 60 or 65 should expect to do so.

Another reason for lengthening the period of paid work is the promise of extended independent living arrangements for elders. Healthy people who are earning full or near-full wages should continue to live independently, as independent living *can be* a tremendous benefit for elders specifically and for society generally. I say "can be" because, again, experiences vary. Frail elders who are forced to age "in place," in their hard-to-keep-up homes that are remote from services, do not make good candidates for extended work lives for the purpose of maintaining a presence in their old neighborhood home. On the other hand, older workers who are fortunate to enjoy good physical and mental health can continue to contribute to their own and to others' benefit by maintaining the lives they have constructed.

The "successful aging" proponents argue that people who lead "productive" lives live longer, healthier, happier lives.[35] While this is, of course, an overly general view of aging, I certainly do not mean to say that the converse is always true. Society is enhanced by the participation of elders in intergenerational work systems, and many elders find personally fulfilling roles in such participation. In the course of researching this book, I encountered dozens of bulletin board posts from men and women (mostly men) from the United Kingdom and Ireland who reported negative reactions to having been forced out of their paid work roles at age 60 or younger. Many said they would much prefer being back at work. Although some did mention the opportunity to earn income that would guarantee the lifestyles they had constructed, most listed reasons that were closer to the heart. It was clear that they had come to identify deeply with their worker roles and did not appreciate being tossed aside as a "redundancy."

Recommendations

Because my work focuses on the United States, my recommendations for improving the tech-work environment for elder citizens are most appropriate for Americans. I have brought in some international "best-practice" scenarios when these help to illustrate my position. These are generally from the United Kingdom, where the Age Concern lobby and others have been especially effective on this front and where much action is now taking place. Each national or regional context will have its own needs and projects, but, in the United States, nine solutions can build toward enhancing the work-technology-aging nexus.

Tailor Retirement Systems for Individual Differences

Most economists agree that the world cannot sustain large numbers of early retirements, and critical scholars note that such systems unfairly penalize both working poor elders and the diminishing younger cohorts who are less likely to recoup their investment in old-age supports.[36] One of the reasons that people retire early is their perceived need for greater flexibility and mobility in late life due to personal circumstances, whether it is a desire to winter in warmer climates, be around grandchildren, or simply not have to report to the cubicle every weekday from 9 to 5 after thirty years of having done so. Increasingly, especially for women, early retirement is a means of finding the time to care for an older relative or for grandchildren. The jump from full-time employment to pension collection is severe for many people, both economically and psychologically. Stepped-down participation, in which organizations offer reduced work schedules and greater opportunity for telecommuting, could help both workers and strained economies.

Age Concern England has suggested flexible retirement as one approach. Workers would choose to take full retirement during the decade between ages 60 and 70.[37]

This strategy addresses what has been a problem for many workers, especially men, in more developed nations' workforces. Management disproportionately targets such workers for redundancies during times of downsizing, and they wind up unemployed before leaving the workforce voluntarily because they cannot find replacement work. The British government has reported that such workers typically would prefer to be employed. The government's welfare-to-work program, New Deal, while it should be observed cautiously for its potential to convert welfare benefits-earners into earning paychecks, advocates rational solutions to the problem of unfair redundancies for 50-plus workers. New Deal 50-Plus is working to change employers' attitudes toward older workers through a massive "Age Positive Campaign," which publicizes the potential of older workers and provides economic incentives to both companies and workers for getting elders retrained and employed. In addition, the government has banned age discrimination in employment, effective in 2006. In the meantime, the rhetoric of "Age Positive" is encouraging employers to curtail such practices as age limits in job advertisements, which have been *de rigeur*.[38]

Perhaps because it is coming comparatively late to the antidiscrimination table with respect to the United States, the United Kingdom has been able to introduce more muscle into its policies and rhetoric on aging and employment. The newness of the law invites an explicit rhetoric that government and advocacy groups are able to insert into dialogue with both employers and workers. Similar pushes are occurring in some other European nations, such as Finland; in others, however, such as Ireland, the national conversation has lagged, and older workers continue to find themselves forced out of a job at age 60.

Policymakers must be careful not to eliminate pension availability indiscriminately. A majority of men over 60 may not need their pensions, but three out of four of the world's women would fall well below the poverty line without theirs.[39] A capricious approach to elongating work life, without attention to individual needs and differences, would further imperil elders who have already been made to feel vulnerable by the dominant trope of "productive aging," a scholarly construct made popular in media discourses that preach the values of dedication to work and vitality in order to achieve "success" in old age. Such expectations devalue other meaningful pursuits of elders, such as relational family experiences, and unfairly hold marginalized elders accountable for their own "failures."[40]

Even with "productive roles," diverse possibilities beyond the world of paid work call elders into service. Retirement, for many people, represents an opportunity to put dull jobs behind them and take up more meaningful roles in the voluntary sector, where they perceive themselves to meet new people, do important work, learn new skills, and have an important position in their community.[41]

Elder-Friendly Policies

Governments should proactively encourage employment sectors to be elder friendly, through such means as providing special benefits for self-employment of elders, discouragement of age-limited hints on job postings, and rewards for retention, re-education, and promotion of longtime workers.

The rise of the service sector and the information economy means that escalating hours of work can be performed through home telecommuting, which large numbers of older workers favor.[42]

Educators and opinion leaders of management professionals should encourage dialogic sharing of skill sets between older workers with significant job exposure and younger workers with updated systems knowledge. This can be achieved by a greater emphasis on collaborative work practices more generally, wherein team members are socialized to value fellow learning-community members' strengths and are enabled to work flexibly. Such a shift will be especially salient for older workers themselves, who may still be conditioned to function in a hierarchical and competitive work universe. Contemporary management models that prize cooperation are essential to such a formula.

In many if not most countries, it is legal to discriminate in hiring on the basis of age. In the United States, where only in certain jobs (such as commercial airline pilots) may employers require a maximum hiring or retirement age, discrimination practices persist but are more insidious. Employment recruitment ads that mention "new college graduates," for example, effectively weed out most older workers.

Make the Educational System Nondiscriminatory
In the United States and most other countries, education more often than not means training the young. When politicians debate education budgets, models, and practices in the United States, they generally privilege K–12 and traditional four-year college degrees. Increasingly, especially in the recent slow economy, rhetoric around "retraining" and "lifelong learning" is earning legitimacy, but few models have found wide economic backing from government or corporate institutions. It is time for educational logics to abandon the bias that people attaining undergraduate degrees have become finished "products" of the educational system. "Lifelong learning" can be an antiquated pejorative that conjures notions of academic busy work for retirees. "Learning" must be tailored for everyone, at every stage. Business models that practice social amortization of learning investment by employees deprives society of the rich contributions citizens can make from various moments and contexts in life.

In the United Kingdom, Age Concern England found that just 7 percent of training budgets is spent on people over the age of 40.[43] Governments should ensure lifelong learning protections for people of all age groups, through workplace policies, college grants, and federal leave acts. Just as universities and a few corporations extend sabbatical credit for several years of service, organizations of a certain size should be required to do so, with flexible options for both employer and employee. For example, professors at many universities are able to take a sabbatical on a seven-year cycle. One semester comes at full pay, the second (optional) at half. Sabbaticals must be approved by central administration according to the array of university needs. Giving senior workers similar opportunities to retool, replenish their enthusiasm, and pursue special projects would create more enthusiastic and creative workforces.

Academics Should Begin at Home
We are good at granting sabbaticals but poor at other systems for renewing the resource of senior faculty. How many gripe sessions among college faculty and administrators involve criticism of "deadwood" faculty who do little to earn their keep?[44] Academics believe they hold few cards for solving the "senior faculty-itis" dilemma, short of early retirement buyouts designed to make room for new junior faculty, who are cheaper to support and who are eligible for investment by administrators. It is, in fact, new faculty for whom most professional development systems are geared.

Two-tier reward systems that give senior faculty a choice between "pure teaching and service" and more research-inclusive tracks are one alternative that works in some universities.[45] Faculty who choose to abandon their research programs after receiving tenure can maintain a respectable performance by becoming master teachers. Smart administrators make such promises possible by heavily investing in these senior faculty's renewal processes, through teaching-focused sabbaticals, instructional budgets, and effective professional development programs, both in-house and external. Many scholars of higher education have followed the lead of Ernest Boyer in suggesting remodeling of our house, with its research, teaching, and service wings, so that "scholars of teaching" might be encouraged to emerge, able to contribute in lively ways to the mission of learning and not be punished for choosing to turn away from the research trajectories that led them to tenure.[46]

Senior colleagues heavily involved in research programs should also receive validation and reward. It is tempting for administrators to hold the door open for senior faculty, with their often-high salaries and benefits costs, to make way for larger numbers of less expensive and more up-to-date junior faculty. Some of these senior faculty do make a lot of money, but, as someone serving in an administrative function at my university, I will say that some of them deserve these rewards and more. Academics who adapt to a dynamic research environment and maintain meaningful and active programs of research into their 60s and 70s should be held dear.

Governments Must Stop Discriminatory Practices Against Older Workers

As the people whose words were recorded in this book show repeatedly, weak measures such as the U.S. Age Discrimination in Employment Act do not constitute real insurance for older workers against age discrimination, especially in its most insidious forms. Stronger legislation and enforcement are necessary. Governments must demonstrate that they are substantively interested in the elimination of ageist practices. In the United States, workers file about 17,000 cases per year claiming age discrimination with the Equal Employment Opportunity Commission, but this agency is able to answer only a few "strategically targeted" cases.[47] Clearly, the courts and advocacy agencies are unable to solve the problem through investigation and testimony.

Older workers are notoriously overrepresented in workforce eliminations in many countries. Women, particularly, are vulnerable due to their frequent lower mandatory retirement ages.[48] In many countries, such as the United Kingdom and Canada, downsized older workers drop out of the labor market due to the unlikelihood of finding suitable replacement work.[49] Frequent practices include compulsory retirement, maximum recruitment age, and age limits on access to training.[50] Beyond such overt and legal schemes, however, tacit discrimination is also likely. Even in countries where it is illegal to discriminate on the basis of age, such as the United States, ageist practices persist. How many older workers are considered for plum repositioning in the dynamic, technology-driven environments of America's large businesses? Forty-plus-year-old Baby Boomers might be perceived as "trainable" and "promotable," but it is unlikely that managers think of silver-haired 60-year-olds when they are preparing to tap the talent pool for the new star of their division. The decreasing dominance of trade unions in U.S. societies abets this lack of attention to older workers' career trajectories in technology-oriented settings.

The United States and other nations should pay close attention, through their legislation and policy practices, to the ILO's Older Workers Recommendation (1980), which Alexander Samorodov cites as a means of encouraging equality of opportunity and treatment for older workers.[51] The recommendation promotes access to vocational guidance and placement services, equal opportunity for employment of choice, full access to vocational training (and retraining) facilities, paid education leave for retraining, and equitable distribution of tasks and benefits.[52] Samorodov suggests as one possible avenue a proactive program of retraining older workers for specific vacancies in the labor market to maximize their employability. He presents a rationale for going the extra mile with older workers, arguing against the logic that older workers take longer to train and are often less oriented toward career development in the latter years of their work lives. Instead, Samorodov cites the general superior work habits and loyalty of older workers, the cost-effectiveness of training a worker with a good track record, the opportunity for a company to enhance its image through a display of political capital (elder employee as poster child of an aging society), and investment in older workers as trainers themselves.[53]

Images of Aging Workers: Representations Must Shift Toward Inclusiveness

Lisa Nakamura has found a synthetic and hypocritical inclusiveness in such television commercials as those that made up Microsoft's "Where Do You Want to Go Today" and MCI's "Anthem" campaigns. These commercials—and the magazine advertisements that accompanied them—construct a bland rainbow of ethnicity, age, and gender.[54] In other words, it is possible and even efficient marketing for some politically potent symbol systems and theories—such as multiculturalism—to become popular expressions as they get tapped into service in order to link business with progressive ideals. Just as "Julia," an African American nurse, got her own sitcom in the 1960s and practically every prime-time fiction TV show now has a prominent gay character, the new-media companies similarly embrace the myth of a harmonic, multigenerational, multicultural world.

Certainly, marketers have awoken to the dawn of the "gray market," as shown in Chapter 7. Magazine pages that five years ago had scant if any representation of aging workers now have them sprinkled in along with those of a multiracial, multigendered workforce. Entertainment images, however, lag considerably. More films along the lines of *Space Cowboys*, as discussed in Chapter 6, demonstrating the full range of experiences of aging technology workers, including their successes and challenges, would help this project.

Although policymakers have no direct hand in the work of information and entertainment production, at least in the United States, educators play a significant indirect role. Communication and media arts educators should tactically labor to sensitize future practitioners to the cultural benefits of recognizing the multifaceted experiences of an aging society. At the very least, it is time to orient journalists to the need for seeking accounts of the lived experience of aging workers. For example, the lead story on the business pages of the *New York Times* on July 12, 2002, shocked me. A three-column color photograph of a well-coiffed, well-heeled, 51-year-old Rose Marie Bravo, chief executive of Burberry fashions, is positioned over the headline "Burberry Finds a Fountain of Youth." The story, written by two female business reporters, begins: "Five years ago, most customers at Burberry qualified for senior citizen movie discounts; today most customers might actually be able to stay awake for the picture." The lengthy report goes on to celebrate Burberry's rising stock after its sound pursuit of a more youthful audience.[55] The generous publicity must have pleased Ms. Bravo, but I doubt she could have applauded the nasty lead.

Stop Retrofitting Physical and Social Architecture to Shoehorn Disabled Citizens into a World for "Normals"
Gerard Goggin and Christopher Newell analyzed the development of Australia's telecommunications network:

> It became evident that the thorough changes in communications and media were having enormous effects on the lives of people with disabilities. Much to our frequent dismay, such transformations were given little attention by governments, regulators, and corporations that dominated the shaping of new technologies. People with disabilities were left off networks (old and new), were a problem and liability to be managed rather than accorded priorities as customers, were excluded or marginalized as citizens in the eyes of governments, were constantly reminded how "special" their needs were, while the needs of the majority were unremarked.[56]

Goggin and Newell go on to argue that disabled and elderly citizens should be included, not merely consulted, in the construction of systems at the outset of planning, and such systems should spring from democratic ideologies that will not somehow result in the further marginalization of these groups.

A European-based policy advocacy group observes that earnest investment in younger technology users today will pay off when those users age:

> Many of today's users of Information and Communications Technology (ICT) will be the older people of 2020. They will have lived with ICT as an integral part of their lives for many years. However, most of these people will be experiencing minor impairments (in vision, hearing, manual dexterity) as a result of the normal ageing process, which affect how they are able to interact with ICT-based products or services.... Mainstream products and services will need to be inclusively designed, and capable of being used by older people, without stigma or perception of disability.[57]

The panel stresses that ICT must be customer led, not technology driven, and that government should take the lead in funding research that will place inclusiveness, not retrofit, out front in technological design.[58]

ICT designers should begin with questions about how *all* users would like to employ their technologies, before some "normal" prototype is drawn and in need of re-engineered interfaces for users who do not measure up to the dominant system. For example, what if keyboards and mice had been ergonomically designed in the first place, rather than being supplemented with designs especially for clumsy arthritic fingers, underscoring the stigma of old age in the workplace?

New Ways of Envisioning "Learning Communities" Can Ensure Intergenerational Successes for the Work Arena

Scholars of higher education have started to borrow from management literature's "learning organizations" and "learning communities" paradigm, popularized in the 1990s by guru Peter Senge and his followers. In short, learning communities practice lifelong learning principles in helping their workplaces adapt to changing environment and maintain momentum for success.[59] Education scholars have moved learning communities into diverse contexts for study, including classroom learning communities, residential (dormitory) learning communities, and service-learning communities.[60]

Institutions of higher learning should find imaginative means to make learning communities intergenerational. Service-learning communities are an obvious choice, wherein a team of students enters an environment that includes elders and engages with the group in dialogue for some beneficial outcome. This might involve the adoption of technology in the community, but the curriculum must protect against heavy handedness on the part of the academic "white knights." Instead, a class with an enrollment of tech-savvy students might approach a community that includes elders with the question, "What do you want technology to do for you?" and listen to the response, then codesign a project that will work beneficially, consulting with the elders along the way. Perhaps the project will result in a Web page, a DVD, or an automated telephone-answering program. Such a result is less important than that it will be designed in a community of learners, with all parties learning about technology intergenerationally.

A more adventurous project is the residential learning community. Early-retired and between-engagements professionals in diverse arenas of business are available to serve as consultants in a broad marketplace. Why not bring them to the themed-dormitory learning communities that are now being introduced on college campuses (no doubt as a hedge against enrollment flows to distance-education providers)? Visiting professionals make important contributions in classrooms and faculty colloquia, but they can have even more meaningful dialogues with undergraduate students who are "living" their very study interests in residential communities.

Academics Must Stop Ignoring Age and Aging

It is cumbersome and perhaps unpleasant to undertake the study of both aging as a phenomenon and intergenerational relationships as situated in society, yet it is essential. If intergenerational collaboration is to result, Margaret Simey has argued, researchers, educators, and policymakers must find means of intensifying the bonds between the generations.[61] Miriam Bernard and Judith Phillips press further, for the adoption of an intergencrational lifecourse perspective to battle ageism through both academics and policy.[62] A successful example that Bernard and Phillips cite is the Gray Panthers, whose Web site *www.graypanthers.org*, promotes intergenerational work and understanding for the elimination of ageist attitudes in society.

Critical gerontology must find more places outside the halls of social work and health care and across the campus, in general-education courses on such topics as human and mediated communication, women's studies, political science, economics, and geography. The questions of critical gerontology and intergenerational relations should not be merely relegated to the "diversity perspective" course but woven across the curriculum to better prepare students for adult life in a society of aging workers and other elders. Through such means, academics can perform the important work of putting "a human face—and a human body and spirit—on ageing and growing old."[63]

Feminists have argued that technology belongs at the center of current policymaking on age and work, because of the following:

1. Technology is intergenerational, connecting or dividing generations in work and other aspects of everyday life.
2. Technology is pervasive, with the potential for solving problems associated with aging or creating and perpetuating such difficulties.

3. Technology is potentially facilitative and empowering for workers, citizens, and consumers (variations on a subjective experience) in their attempts to function seamlessly in society.[64]

Yet scholarship on technology retains its teenaged crush on youth, with a proliferation of study on young people's identity formations, young transsexuals' virtual gender bending, and the like. Such scholarship is welcome in an environment in which the diversity of youth is ever more pronounced and the human experience of time and space so dramatic. The academy should do its best to encourage more research on phenomena surrounding the experience of aging, especially since aging has, it seems, come into style. As I previously noted, anthropologist Margaret Clark noted almost four decades ago that few of her peers undertook aging studies because they saw the subject as a "morbid preoccupation—an unhealthy concern, somewhat akin to necrophilia."[65] It was as though, Clark felt, the anthropologist might fear the contraction of decay and death by peering too closely into the Pandora's box of age. But now that age undeniably has come to us—to the world, generally, and to the academy, specifically, *where one-third of faculty members are over age 55*[66]—we had better get to work.

The New Age

To conclude, I am asserting the need for us all to assume some collective sense of responsibility for recreating social environments that will accommodate the unprecedented complexities of intergenerational living in today's world. It is insufficient for aging Baby Boomers to learn new models for preretirement and for retirement, although that certainly is necessary. Likewise, it is impractical to expect that individual employers, faculty committees, and public authorities can reverse the challenging circumstances of ageist practices in our contemporary environment. But such reversals must begin, and they must begin multilaterally. Chambers of commerce, labor unions, creative guilds, and nonprofit boards must take up these themes, as must physicians' groups, purchasing consortiums, industrial engineers, moviegoers, and voters.

To say that we are all implicated in the many outcomes of such a public dialogue is to acknowledge both that we play roles in the deliberative processes that the dialogue touches and also that we are personally situated in the intergenerational morass that encompasses our society. For example, we are either a member of an older generation or we have relationships to elders, perhaps both. Likewise, if we are old, we have relationships to the young. In the case of Baby Boomers, we most often are in the middle, relating to both old and young and enduring circumstances that pinpoint us at a crossroads between age and youth ourselves.

Although I can hope that all of us—Matures, Boomers, Xers, and Millennials—can participate in the sort of democratic conversations and upheavals that are necessary to effect progressive change in the intergenerational workplace, I will be both surprised and delighted if some sorts of voices are not heard over others. This book, in tracing the contours of age, work, and technology together, has acknowledged the persistence of such dividing structures as gender, race, ethnicity, and social class. It is essential for those of us who would act in progressive ways to find means of including the voices of those who might otherwise be absent from the dialogue.

Notes

Chapter 1

1. Simone de Beauvoir, *The Coming of Age*, trans. Patrick O'Brian. (New York: W.W. Norton, 1996).
2. Susannah Fox, *Wired Seniors*, Washington, D.C.: Pew Internet and American Life Project, *www.pewinternet.org*, September 2001.
3. U.S. Bureau of the Census, *Sixty-Five Plus in the United States, www.census.gov/publications/ socdemo/statbriefs/agebrief.html,* accessed January 2003.
4. U.S. Bureau of the Census, *National Vital Statistics Report* 47:28, December 13, 1999.
5. Richard Disney, *Can We Afford to Grow Older? A Perspective on the Economics of Aging* (Cambridge, Mass.: MIT Press, 1996).
6. Ibid.
7. Karen E. Riggs, *Mature Audiences: Television in the Lives of Elders* (New Brunswick, N.J.: Rutgers University Press, 1998).
8. Paul Wallace, *Agequake: Riding the Demographic Rollercoaster Shaking Business, Finance and Our World* (London: Nicholas Brealey Publishing, 1999).
9. Tom Brokaw, *The Greatest Generation* (New York: Delta Books, 2001); Jimmy Carter, *The Virtues of Aging* (New York: Ballantine Books, 1998); and Betty Freidan, *The Fountain of Age* (New York: Simon & Schuster, 1993).
10. U.S. Bureau of the Census, *National Vital Statistics Report.*
11. U.S. Bureau of the Census, *Sixty-Five Plus in the United States.*
12. David A. Wise, ed., *Advances in the Economics of Aging* (Chicago: University of Chicago Press, 1996).
13. Ibid.
14. Wallace, *Agequake.*
15. Disney, *Can We Afford to Grow Older?*
16. Angus Deaton and Christina Paxson, "Health, Income, and Inequality over the Life Cycle," in *Frontiers in the Economics of Aging*, ed. David A. Wise (Chicago: University of Chicago Press, 1998), pp. 45–62.
17. Harry R. Moody, *Abundance of Life: Human Development Policies for an Aging Society* (New York: Columbia University Press, 1988).
18. Ibid.
19. Rose Rubin and Michael O. Nieswiadomy, *Expenditures of Older Americans* (Westport, Conn.: Praeger, 1997).
20. Robert N. Butler, *Why Survive?: Being Old in America* (New York: Harper, 1975).
21. Bill Bytheway, "Youthfulness and Agelessness: A Comment," *Ageing and Society* 20, no. 6 (November 2000): 781–89.
22. Julie Ann McMullin and Victor W. Marshall, "Ageism, Age Relations, and Garment Industry Work in Montreal," *The Gerontologist* 41, no. 1: 111–122.
23. Robert N. Butler, "Introduction," in *Productive Aging: Enhancing Vitality in Later Life*, ed. Robert N. Butler and Herbert P. Gleason (New York: Springer Publishing Co., 1985), 3.
24. Butler and Gleason, *Productive Aging.*
25. Carroll L. Estes and Associates, *Social Policy and Aging: A Critical Perspective* (Thousand Oaks, Calif.: Sage, 2001).
26. Chris Carlsson, "The Shape of Truth to Come: New Media and Knowledge," in *Resisting the Virtual Life: The Culture and Politics of Information*, ed. James Brook and Iain Boal (San Francisco: City Lights, 1995), 238.
27. Pippa Norris, *Digital Divide: Civic Engagement, Information Poverty and the Internet Worldwide* (London: Cambridge University Press, 2001).
28. Ibid.

29. Neil Charness, Denise C. Park, and Bernhard A. Sabel, eds., *Communication, Technology and Aging* (New York: Springer Publishing Co., 2001).
30. Such critiques proceed from both structuralist and poststructuralist/postmodern positions, with theorists acknowledging that the technological forces of patriarchal capitalism both contain some element of stability and are subject to multiple and complex revisions.
31. Webster documents the significance of women's job losses due to automation, the ways in which sexual and racial division at various levels in the workplace may route corporate decisionmakers toward cheaper structures of employment, and how new technologies have resulted in a spiraling "feminization" of some jobs that were formerly held by skilled (male) workers. Juliet Webster, *Shaping Women's Work: Gender, Employment and Information Technology* (London: Longman, 1996).
32. Cynthia Cockburn and Ruza Fürst-Dilic, eds., *Bringing Technology Home: Gender and Technology in a Changing Europe* (Buckingham, UK, and Philadelphia: Open University Press, 1994).
33. Carol Ann Meares, John F. Sargeant Jr., et al., *The Digital Work Force: Building Infotech Skills at the Speed of Innovation* (Washington, D.C.: U.S. Department of Commerce Technology Administration, Office of Technology Policy, 1999).
34. Peter G. Peterson, *Gray Dawn: How the Coming Age Wave Will Transform America—and the World* (New York: Random House, 1999).
35. Ibid.
36. Wallace, *Agequake*.
37. Larry Naylor has grappled with this question similarly in his edited volume, *Problems and Issues of Diversity in the United States* (Westport, Conn.: Bergin and Garvey, 1999). Naylor notes that scholarship is increasingly consumed with issues of diversity but pays only scant attention to elders.
38. Many scholars have addressed the lack of old age as a salient identification for most people. See, for example, Andrew Blaikie, *Ageing and Popular Culture* (New York: Cambridge University Press, 1999).
39. Margaret Clark, "The Anthropology of Aging, A New Area for Studies of Culture and Personality," *The Gerontologist* 7: 55–64, p. 55.
40. Kathleen Woodward, *Aging and Its Discontents: Freud and Other Fictions* (Bloomington: Indiana University Press, 1991); and Kathleen Woodward, ed., *Figuring Age: Women, Bodies, Generations* (Bloomington: Indiana University Press, 1999).
41. "An Aging World," U.S. Bureau of the Census, November 2001.
42. S.M. Keigher, "The Limits of Consumer Directed Care as Public Policy in an Aging Society," *Canadian Journal on Aging* 18, no. 2 (1999): 182–210.
43. See, for example, the Xer backlash literature recorded in Richard D. Thau and Jay S. Heflin, eds., *Generations Apart: Xers Vs. Boomers Vs. the Elderly* (Amherst, N.Y.: Prometheus Books, 1997).
44. See, for example, J. Walker Smith and Ann Clurman, *Rocking the Ages: The Yankelovich Report on Generational Marketing* (New York: HarperBusiness, 1997).
45. Drew Leder, *The Absent Body* (Chicago: University of Chicago Press, 1990).
46. See Thuy-Phuong Do and Patricia Geist, "Embodiment and Dis-Embodiment: Identity Trans-Formation of Persons with Physical Disabilities," in *Handbook of Communication and People with Disabilities: Research and Application,* ed. Dawn O. Braithwaite and Teresa L. Thompson (Mahwah, N.J.: Lawreance A. Erlbaum, 2000), 54.
47. Webster, *Shaping Women's Work*, p. 118, invokes the role of secretaries as "office wives."
48. Advanced age, like all other cultural categories, receives its meaning from context. Pia Kontos, in arguing for a phenomenological approach to critical gerontology, says that in order to understand aging and what it means to inhabit an aging body, we must acknowledge the "plasticity of biology and its interdependence with culture." The clash of cultural categories that inform the experiences of aging must be taken into account in order to understand the complexities of the social milieux. Place, as Kontos argues, is integral to the experience and construction of age and the aging body. Kontos focuses on home as the site of much of that experience and construction, but I am arguing that the workplace, wherein so much of our social experience unfolds, is of near equal importance and for some is of greater importance. See Pia Kontos, "Resisting Institutionalization: Constructing Old Age and Negotiating

Home," in *Aging and Everyday Life*, ed. Jaber F. Gubrium and James A. Holstein (Malden, MA: Blackwell Publishers, 2000), 259.
49. Matilda W. Riley, "Aging and Cohort Succession: Interpretations and Misinterpretations," *Public Opinion Quarterly*, 37, no. 1 (1973): 35–49.
50. U.S. Bureau of Labor Statistics, www.bls.gov/iif/oshcdnew.htm.

Chapter 2

1. National Center for Policy Analysis, "Baby Boomers Now Encountering Layoffs," August 16, 2001, www.ncpa.org/~ncpa/pd/economy/pd081601c.html
2. See, for example, Adam Geller, "Job Layoffs Linked to Age, Boomers Say," *South Jersey News/Courier Post Online*, July 1, 2001, www.southjerseynews.com/issues/july/b07010d.htm
3. Katharine Mieszkowski, "The Age of Overwork," *Salon.com*, March 1, 2001, http://archive.salon.com/tech/feature/2001/03/01/white_collar_sweatshop/index.html
4. Carol Ann Meares, John F. Sargent Jr., et al., *The Digital Workforce: Building Infotech Skills at the Speed of Innovation*, (Washington, D.C.: U.S. Department of Commerce Technology Administration, Office of Technology Policy, 1999); and Neal Weinberg, "Help Wanted: Older Workers Need Not Apply," *Network World*, posted on CN Interactive on September 14, 1998.
5. http://www.aarp.org/working_options/
6. Apropos the media's translation of social problems into stories of individual problem solving is Angela McRobbie's influential critique of the British magazine for teenage girls, *Jackie*. See Angela McRobbie, "*Jackie*: An Ideology of Adolescent Femininity," Stencilled Occasional Paper, *Women Series SP No. 53* (Birmingham: Centre for Contemporary Cultural Studies, 1982).
7. www.careerpath.com/tutorial/Candidates/smarter
8. Ibid.
9. Ibid.
10. See, for example, Rob Shields, ed.,*Cultures of Internet: Virtual Spaces, Real Histories, Living Bodies* (London: Sage, 1996); Nancy Baym, "Interpreting Soap Operas and Creating Community: Inside an Electronic Fan Culture," in *Culture of the Internet*, ed. Sara Kiesler (Mahway, N.J.: Lawrence Erlbaum Associates, 1997); and Sherry Turkle, "Constructions and Reconstructions of Self in Virtual Reality: Playing in the MUDS," in *Culture of the Internet*, ed. Kiesler.
11. Erving Goffman, *The Presentation of Self in Everyday Life* (New York: Anchor/Doubleday, 1959).
12. See Shawn P. Wilbur, "An Archaeology of Cyberspaces: Virtuality, Community, Identity," in *The Cybercultures Reader*, ed. David Bell and Barbara M. Kennedy (London: Routledge, 2000), 45–55.
13. Marjorie Kibby, "Babes on the Web: Sex, Identity and the Home Page," *Media International Australia* 84 (May 1997): 39–45.
14. Jamie M. Poster, "Trouble, Pleasure, and Tactics: Anonymity and Identity in a Lesbian Chat Room," in *Women and Everyday Uses of the Internet*, Mia Consalvo and Susanna Paasonen, eds. (New York: Peter Lang, 2002), 230–353.
15. Steve Jones has addressed the potentially evanescent nature of CMC communities in his excellent essay on virtual community. See Steven G. Jones, "Understanding Community in the Information Age," in *Cybersociety*, Steven G. Jones, ed. (Thousand Oaks, Calif.: Sage Publications, 1995), 10–35.
16. See, for example, Nancy Baym, "The Emergence of On-Line Community," in *Cybersociety 2.0: Revisiting Computer-Mediated Communication and Community*, Steven G. Jones, ed. (Thousand Oaks, Calif.: Sage, 1998), 35–68.
17. Mike Featherstone, "Post-Bodies, Aging and Virtual Reality," in *The Cybercultures Reader*, ed. Bell and Kennedy (London: Routledge, 2000), 612.
18. Mia Consalvo, "Marketing to Women: Finding the Female Gaming Audience" (paper presented at Game Developers Conference, San Jose, Calif., March 2003).
19. This practice is similar to the romance novel readers of Janice Radway's research who escaped from everyday troubles but did not face up to their life challenges. See Janice Radway, *Reading*

the Romance: Women, Patriarchy, & Popular Literature (Chapel Hill: University of North Carolina Press, 1984).
20. This phenomenon has been well documented by such scholars as Nancy Baym, *Tune In, Long On: Soaps, Fandom, and Online Community* (Thousand Oaks, Calif.: Sage, 2000).
21. Alluquere Roseanne Stone, "Will the Real Body Please Stand Up? Boundary Stories About Virtual Cultures," in *Cyberspace: First Steps*, ed. Michael Benedikt, Cambridge, Mass.: MIT Press, 1991).
22. Peter Lovelock and John Ure, "The New Economy: Internet Telecommunications and Electronic Commerce," in *Handbook of New Media: Social Shaping and Consequences of ICTs*, ed. Leah A. Lievrouw and Sonia Livingstone, (London: Sage, 2002), 363.
23. Tara M. Kachgal, "Goddess Worship: Commodified Feminism and Spirituality on NIKEgoddess.com" (paper presented at Association for Education in Journalism and Mass Communication, Miami, August 2002).
24. These findings echo the conclusions of Susan Faludi, *Stiffed: The Betrayal of the American Man* (New York: Perennial, 1999).
25. Discussions of netiquette violations can be found in Margaret L. McLaughlin, Kerry K. Osborne, and Christine B. Smith, "Standards of Conduct on Usenet," in *Cybersociety*, ed. Jones; and Baym, *Tune In, Long On* (1995).
26. Murray Edelman, *Constructing the Political Spectacle* (Chicago: University of Chicago Press, 1988); Matt Wray and Annalee Newitz, eds., *White Trash: Race and Class in America* (New York: Routledge, 1997).
27. I have not corrected typographical errors in the posts.
28. The issue of Internet users' anonymity and its relationship to credibility and authority of the information shared there is undertaken by several authors, including Laura Gurak, *Cyberliteracy: Navigating the Internet with Awareness* (New Haven, Conn.: Yale University Press, 2001); and Byron Burkhalter, "Reading Race Online: Discovering Racial Identity in Usenet Discussions," in *Communities in Cyberspace*, ed. Marc A. Smith and Peter Kollock (London: Routledge, 1999).
29. Roger S. Slack and Robin A. Williams, "On Community in the 'Information Age'," *New Media and Society*, 2, no. 3(2000): 313–34.
30. Jennifer M. Tiernan, "Women Veterans and the Net: Using Internet Technology to Network and Reconnect," in *Women and New Media*, ed. Consalvo and Paasonen, 211–27.
31. For analysis of the role of gossip in women's informal networks, see Patricia Spacks, "In Praise of Gossip," *Hudson Review*, 35(1982): 19–38; Mary Field Belenky, Blythe McVicker Clinchy, Nancy Rule Goldberger, and Jill Mattuck Tarule, *Women's Ways of Knowing: The Development of Self, Voice, and Mind* (New York: Basic Books, 1986); and Mary Ellen Brown, "Motley Moments: Soap Opera, Carnival, Gossip and the Power of the Utterance," in *Television and Women's Culture: The Politics of the Popular* ed. Mary Ellen Brown, (London: Sage, 1990).
32. Tara McPherson, "I'll Take My Stand in Dixie-Net: White Guys, the South and Cyberspace," in *Race in Cyberspace*, ed. Beth E. Koko, Lisa Nakamura, and Gilbert B. Rodman (New York: Routledge, 2000), 117–31.

Chapter 3

1. Simone de Beauvoir, *Coming of Age*, trans. Patrick O'Brian (New York: W.W. Norton, 1972), p. 5.
2. See, for example, Julie Ann McMullin and Victor W. Marshall, "Ageism, Age Relations, and Garment Industry Work in Montreal," *The Gerontologist*, 41, no. 1(2001): 111–22.
3. Adriana Vandenhuevel, "Mature Age Workers: Are They a Disadvantaged Group in the Labour Market?" *Australian Bulletin of Labour* 25, no. 1 (1999): 11–12. See also Ranjita Misra and Bhagaban Panigrahi, "Effects of Age and Attitudes Towards Working Women," *International Journal of Manpower* 17, no. 2 (1996): 3–17.
4. A variety of studies done by social scientists has established that, generally speaking, the older that workers are, the more likely they are to feel less comfort, efficacy, and control over computers. See Sara J. Czaja and Joseph Sharit, "Age Differences in Attitudes Toward Computers," *Journal of Gerontology* 53B, no. 5 (1998): 329–40. See also J.C. Marquie, B. Thon,

and B. Baracat, "Age Influence on Attitudes of Office Workers Faced with New Computerized Technology," *Applied Ergonomics* 25 (1994): 130–42.
5. Juliet Webster, *Shaping Women's Work: Gender, Employment and Information Technology* (London and New York: Longman, 1996).
6. Ibid.
7. The images of these hourly workers contributes to the marketing of Harley's mystique. This practice contrasts with catalogues from some other companies. For example, L.L. Bean, the Maine manufacturer of rugged outdoor gear, used employee models in its catalogs until it went upscale and hired professionals to sell a more yuppy-friendly image. Nancy's Notions, a Wisconsin sewing-supply catalog business with a national consumer base, has used its employees as catalog models for many years. Presumably, this company's home sewing constituency is unoffended by the modest production values to which its "models" contribute.
8. Tacit cues from up and down the corporate hierarchy encourage Harley-Davidson ownership. Casual conversation generally turns to questions about what kind of bike one owns. Managers learn quickly that owning and riding one's own Harley-Davidson is as valuable as taking up golf. One of the managers we interviewed had taken up motorcycling only since she had been promoted to a high-visibility position in the company. Owning the product provides not only a social lubricant among members of the management team but a crucial point of contact between the management and labor class. Most of the blue-collar female employees own Harley-Davidsons, and the higher that women in corporate headquarters go up the management ladder, the more likely that they own them, too.
9. See, for example, Laura L. Bierema, "How Executive Women Learn Corporate Culture," *Human Resource Development Quarterly* 7, no. 2 (1996): 145–64.
10. Many workers felt edgy about the prospect of disclosing personal views about work in the corporate setting, although the project was marketed as separate from the company and identities were to be confidential.
11. Webster, *Shaping Women's Work*.
12. Donna Haraway, *Simians, Cyborgs, and Women: The Reinvention of Nature* (New York: Routledge, 1991).
13. Women remain overrepresented in the so-called pink-collar ghetto of low-paid office work, while executives remain dominantly male, and this gender dynamic has had a continued impact on the distribution of computer resources as well as division of functions. Such practices as passing "hand-me-down" computers to female support staff indicate the broader social conditions of patriarchy. For further explanation, see Webster, *Shaping Women's Work*.
14. Cynthia Cockburn, "The Circuit of Technology: Gender, Identity, and Power," in *Electronic Media and Technoculture*, ed. John Thornton Caldwell (New Brunswick, N.J.: Rutgers University Press, 2000), 209.
15. Ellen Seiter, "Television and the Internet," in *Electronic Media and Technoculture*, ed. Caldwell, 227–43.
16. This anxiety about the lack of time available extended to not having time to take a few minutes periodically to examine the Harley-Davidson corporate Web site, and thus feel in control of information one should have about company business. Some of the women felt this was an obligation they felt they should but could not fulfill.
17. The women who work at the manufacturing plants are quite conscious of a class divide between themselves and the primarily male employees of the nearby engineering facility, who are largely professionals. They know that the engineering group workers avoid visiting the comparatively coarse Powertrain Operations plant, for example, and they realize that email has abetted this reticence.
18. Arlie Hochschild, *The Time Bind: When Work Becomes Home and Home Becomes Work* (New York: Metropolitan Books, 1997).
19. At least two participants were planning to buy their first home computers but not until after January 1, 2000. On the whole, the women said they had a few concerns that the Y2K bug would affect their work, and a few said they did not want to invite trouble at home, too.
20. Webster, *Shaping Women?s Work*.
21. See Webster on the pairing of patriarchy and capitalism.
22. Barbara A. Gutek and Laurie Larwood, "Information Technology and Working Women in the USA," in *Women and Information Technology*, ed. Marilyn J. Davidson and Cary L. Cooper (Chichester, England, and New York: John Wiley and Sons, 1987).

23. David E. Nye, *Consuming Power: A Social History of American Energies* (Cambridge, Mass.: MIT Press, 1998).

Chapter 4

1. Jackie Vinson and Amy Lauters helped conduct interviews and collaborated on the paper on which this chapter is based.
2. Bernie Herlihy, "Targeting 50+: Mining the Wealth of an Established Generation," *Direct Marketing* 61, no. 7 (November 1998): 19.
3. Ibid., p. 20.
4. This quote can be found in, among countless other feature articles, a story by Dale English, "Surfin' Seniors: Elders Don't Have to Be Bumped Off the Information Superhighway," *Business First* 15, no. 49 (September 6, 1999): 23–24.
5. Cheryl Russell, "The Haves and the Want-Nots," *American Demographics* 20, no. 4 (April 1998): 10.
6. Approximately one to one-and-a-half-hour interviews were conducted with volunteer participants, identified through access to computer-training programs of Milwaukee Interfaith Programs for the Elderly, Milwaukee Urban League, and Milwaukee Public Library. Generally, interviews were conducted by a student and professor or by two students. In many cases, follow-up telephone interviews were conducted. The women who were interviewed were between 50 and 80 years old.
7. Dan Gebler, "Rethinking E-commerce Gender Demographics," *E-Commerce Times* www.ecommercetimes.com/news/articles2000/001006-2.shtml.
8. Victoria Tauli-Corpuz, "ICTs: Their Impact on Women and Proposals for a Women's Agenda," *Women in Action* 2 (1999): 96.
9. Sue Clegg, "Is Really Computing for Girls?: A Critical Realist Approach to Gender Issues in Computing," in *After Postmodernism: An Introduction to Critical Realism*, ed. G. Potter and J. Lopez (London: Athlone Press, 2000).
10. This logic is supported by the recent ethnographic work done by the principal researcher and a graduate student on Harley-Davidson in Karen E. Riggs and Joette Rockow, "An American (Techno)Legend: Older Women Work to Position Themselves in a Computerized Shop" (paper presented to the Feminist Scholarship Division of the International Communication Association, Acapulco, 2000).
11. Dionne Brand, "Black Women and Work: The Impact of Racially Constructed Gender Roles on the Sexual Division of Labour," *Fireweed* 25 (1987): 28.
12. Sandra S. Butler and Richard A. Weatherley, "Poor Women at Midlife and Categories of Neglect," *Social Work* 37, no. 6 (1992): 510–16.
13. Kathleen M. Roe and Meredith Minkler, "Grandparents Raising Grandchildren: Challenges and Responses," *Generations* 22, no. 4 (1999): 25–32. Roe and Minkler note that the number of children living with grandparents and other relatives grew in the 1980s by 44 percent and reached 4 million children, or 5.5 percent of all children in the United States, living with a grandparent by 1997; one-third of these families did not include a biological parent present.
14. The facts in this paragraph are attributable to research discussed in Roe and Minkler.
15. Roe and Minkler.
16. Joseph G. Altonji, Ulrich Doraszelski, and Lewis Segal, "Black/White Differences in Wealth," *Economic Perspectives* 1 (2000): 38–51.
17. For more on older African American women and job history complications, see Laurie Russel Hatch, "Gender and Work at Midlife and Beyond," *Generations*, 14, no. 3 (1990): 48–53.
18. See Karen E. Riggs, *Mature Audiences: Television in the Lives of Elders* (New Brunswick, N.J.: Rutgers University Press, 1998).
19. Donna Haraway, "A Cyborg Manifesto: Science, Technology and Socialist-Feminism in the Late Twentieth Century," *Simians, Cyborgs, and Women: The Reinvention of Nature* (New York: Routledge, 1991), 155.
20. Ibid., 175.
21. I mean this word in the sense that Michel Foucault uses it in his work on disciplinary technology; see *Discipline and Punish: The Birth of the Prison*, trans. Alan Sheridan

(New York: Random House, 1977). Foucault refers to the architectural centerpiece of a prison, in which guards may maintain surveillance constantly over all prisoners at the periphery, ensuring "the automatic function of power" (p. 201).
22. Neil Postman, *Technopoly: The Surrender of Culture to Technology* (New York: Vintage Books, 1993) p. 9.
23. Erving Goffman, *Stigma: Notes on the Management of Spoiled Identity* (Englewood Cliffs, N.J.: Prentice-Hall, 1963).
24. David T. Mitchell and Sharon L. Snyder, *The Body and Physical Difference: Discourses of Disability* (Ann Arbor: University of Michigan Press, 1997).

Chapter 5

1. Rob Shields, *The Virtual* (London: Routledge, 2003), 114.
2. Ibid., 129–30.
3. John T. Waisanen theorized the virtual world in terms of geometric associations that are felt "on the body." Following the work of Walter Ong, who described the "shifting sensorium" of moving across multiple modes of experience and communication, Waisanen suggested that the incorporation of the Internet and other new media experiences into everyday life has contributed to the experience of consciousness in visual, oral, and literate modes, a complicated knowledge set indeed. See John T. Waisanen, in *Thinking Geometrically: Re-Visioning Space for a Multimodal World*, ed. Jennifer Daryl Slack (New York: Peter Lang, 2002). Also see Walter J. Ong, *The Presence of the Word: Some Prolegomena for Cultural and Religious History* (New Haven, Conn.: Yale University Press, 1967).
4. Manuel Castells, *The Rise of the Network Society*, 2d ed. (Oxford, UK: Blackwell, 2000), 475.
5. Ibid., 476.
6. Ibid.
7. Ibid., 475.
8. Manuel Castells, *The Internet Galaxy: Reflections on the Internet, Business and Society* (Oxford, UK: Oxford University Press, 2001).
9. Ibid., 92.
10. Ibid., 90.
11. Ibid., 91.
12. See Thomas S. Kuhn, *The Structure of Scientific Revolutions*, 2d ed. Chicago: University of Chicago Press, 1970). Regarding the argument that the "Matures" expect an accrual of knowledge rather than ruptures, I am asserting this position with the notion that this generation of Americans largely behaved as if the Great Depression were anomalous to the social and political context of the twentieth century. The enormous confidence wrought by the postwar expansion economy taught them to expect that what they had come to recognize as a century of progress was authentic. That logic was not dislodged until the disruptive narratives associated with the Baby Boom entered the public imagination.
13. Ron Zemke, Claire Raines, and Bob Filipczak, *Generations at Work: Managing the Clash of Veterans, Boomers, Xers, and Nexters in Your Workplace* (New York: Amacom, 2000).
14. Castells, *The Internet Galaxy*, 94.
15. Michel Foucault, *Discipline and Punish: The Birth of the Prison*, trans. Alan Sheridan (New York: Random House, 1977).
16. See, for example, Torry D. Dickinson and Robert K. Schaeffer, *Fast Forward: Work, Gender, and Protest in a Changing World* (Lanham, Md.: Rowan & Littlefield, 2001).
17. During January and February 2003, my research assistant and I posted requests for participation on approximately twenty Web sites, most of which were based in the United States. These included employment advice sites, aging-related sites, "geek" sites (such as www.slashdot.com), and professional organization sites related to information technology, such as engineering societies. We got pitifully few responses, perhaps signifying the overwhelming other "spam-like" messages that people encounter. The responses were amazingly even in terms of gender, generational cohort, income, and my designation of them as either "generic" or "self-programmable."
18. I am not suggesting anything conclusive from these statements, only that the group of people I interviewed revealed these characteristics. It is worth noting, however, whether the lack of

"international" respondents in the generic category resonates with the idea that people who use the Internet "globally" tend to exercise broader perspectives than people who do not.
19. Helen Dennis, *Fourteen Steps in Managing an Aging Workforce* (Lexington, Mass.: Lexington Books/DC Heath, 1988).

Chapter 6

1. Andrew Blaikie, *Ageing and Popular Culture* (Cambridge, UK: Cambridge University Press, 1999).
2. Anthony Leong, "The Movie Business in 1997," www.mediacircus.net/film1997.html, 1998, and www.lucineentertainment.com/invest.html, visited, January 2003.
3. Blaikie, *Ageing and Popular Culture.*
4. Drawing on the discourses of discipline theorized by Michel Foucault, Elaine Graham considers representations of the "post/human" agent, including film, as spaces in which the moral economy or "ontological hygiene" of cyborgs and related others might be worked out anew. For Graham, filmic representations that valorize prosthetic and biomedical progressions of these "post/human" agents both accent the ambiguities of identity in contemporary industrial society and become circulated into popular discourse as "crucibles of production of the very reality we inhabit"—forming, transforming, and maintaining our identities. Representations of characters that have escaped the trappings of the flesh, Graham believes, create possibilities for us to consider as the array of roles for "post/human" actors in society gets negotiated. See Elaine L. Graham, *Representations of the Post/Human: Monsters, Aliens and Others in Popular Culture* (New Brunswick, N.J.: Rutgers University Press, 2002). In the case of aging characters, movies described in this chapter, such as *X-Men, Driving Miss Daisy, Back to the Future,* and *Cocoon,* encourage a revisioning of the aging body in the world, allowable through the development of technology as social prosthesis. I mean to say not that such films encourage us to see the aging body as free of itself through technology. Instead, they promote the possibility that technology allows the aging body to interact in new and complex ways within the environment; joining other scholars of new technology, I stop short of insisting that the self is ever unmoored from the body. See, for example, Beth E. Kolko, Lisa Nakamura, and Gilbert B. Rodman, *Race in Cyberspace* (New York: Routledge, 2000).
5. Simon During, "Popular Culture on a Global Scale: A Challenge for Cultural Studies?" in *The Media Reader: Continuity and Transformation,* ed. Hugh Mackay and Tim O'Sullivan (London: Sage, 1999), 211–22.
6. Vladimir Propp, in *The Morphology of the Folk Tale,* addresses the consigned status of elders in what has come to be known as the fairy tale. See Propp (Bloomington: Indiana University Research Center in Anthropology, Folklore, and Linguistics, 1958).
7. Patricia Mellencamp, *A Fine Romance: Five Ages of Film Feminism* (Philadelphia: Temple University Press, 1995), 165.
8. See, for example, Kathleen Woodward, *Aging and its Discontents: Freud and Other Fictions* (Bloomington: Indiana University Press, 1991), and Tamara K. Hareven, "Changing Images of Aging and the Social Construction of the Life Course," in *Images of Aging: Cultural Representations of Later Life,* ed. Mike Featherstone and Andre Wernick (London: Routledge, 1995), 119–34.
9. Mellencamp, *A Fine Romance.*
10. Vivian Sobchack, "Scary Women: Cinema, Surgery, and Special Effects," in *Figuring Age: Women, Bodies, Generations,* ed. Kathleen Woodward (Bloomington: Indiana University Press, 1999), 200–11.
11. www.amazon.co.uk
12. Likewise, it is with shrewd marketing knowledge that Hollywood cast a young, attractive Sandra Bullock as the centerpiece of *The Net.* No Vanessa Redgrave there.

Chapter 7

1. For details, see William Leiss, Stephen Kline, and Sut Jhally, *Social Communication in Advertising: Persons, Products and Images of Well-Being* (Scarborough, Ontario: Nelson Canada, 1990).

2. Some feminist studies of beauty practices have built on the work of Michel Foucault to explain how women continually construct their bodies in response to regulating discourses. See Sandra Lee Bartky, "Foucault, Femininity and the Modernization of Patriarchal Power," in *Femininity and Domination: Studies in the Phenomenology of Oppression* (New York: Routledge Press, 1990), 63–82; and Susan Bordo, "Anorexia Nervosa: Psychopathology as the Crystallization of Culture," in Inene Diamond and Lee Quinby, eds., *Feminism and Foucault: Reflections on Resistance*, 87–117 (Boston North-eastern University Press, 1988).
3. *A Nation Online: How Americans are Expanding Their Use of the Internet*, Washington, D.C.: February 2002. Significantly, older women, who on the whole lack job exposure to the Internet as compared to older men, are resisting the Internet in greater numbers, and, generally, Americans over 65 remain the most resistant age group, according to the Pew Foundation Project on the Internet and American Life, as reported on NPR, "Study: Senior Women Reluctant to Cross Digital Divide," Weekend Edition, February, 23, 2003.
4. Chris Holmlund, *Impossible Bodies: Femininity and Masculinity at the Movies* (London: Routledge, 2002).
5. Ibid., 19.
6. See, for example, the images discussed by Andrew Blaikie, *Ageing and Popular Culture* (Cambridge, UK: Cambridge University Press, 1999).
7. The news is not completely negative. In February 2003, www.feminist.com, a grassroots activist community, reported this on its Web site: "Women now hold 15.7 percent of corporate officer positions in Fortune 500 companies, up from 12.5 percent in 2000. This is one of the largest increases in the past few years, according to a recent census report by Catalyst, a New York City–based organization that studies women and business trends. The numbers were just 8.7 percent in 1995, when Catalyst first began tracking them."
8. Elizabeth Grosz, *Volatile Bodies: Toward a Corporeal Feminism* (Bloomington: Indiana University Press, 1994), 120.
9. Kathleen Woodward, *Aging and Its Discontents: Freud and Other Fictions* (Bloomington: Indiana University Press, 1991).
10. My continua are hopelessly postmodern. I am asserting that we occupy relative positions at any particular time and in a particular aspect of life, although I will acknowledge that many patterns emerge and start to look familiar in a variety of applications. Here, like Emmanuelle Tulle-Winton, I echo M. Weiss, who argues that we are "multiple and inconsistent embodied selves." Tulle-Winton contends that the lived experience of old bodies is contingent on the structures and narratives that intersect with them. See M. Weiss, "Narratives of Embodiment: The Discursive Formulation of Multiple Bodies, *Semiotica* 118, no. 3/4: 239–60.
11. Michel Foucault, *Discipline and Punish: The Birth of the Prison*, trans. Alan Sheridan (New York: Random House 1977).
12. Emmanuelle Tulle-Winton, "Old Bodies," in *The Body, Culture and Society: An Introduction*, ed. Philip Hancock et al. (Buckingham, UK: Open University Press, 2000). Mike Featherstone and Mike Hepworth, "The Mask of Aging and the Postmodern Life Course," in *The Body: Social Process and Cultural Theory*, ed. Mike Featherstone, Mike Hepworth, and B.S. Turner (London: Sage, 1991); and Kathleen Woodward, *Aging and Its Discontents* (Bloomington: Indiana University Press, 1991).
13. Tulle-Winton, "Old Bodies," p. 77.
14. Mike Featherstone and Andrew Wernick, *Images of Aging: Cultural Representations of Later Life* (London: Routledge, 1995). See also Mike Hepworth, "Positive Aging: What is the Message?" in *The Sociology of Health Promotion: Critical Analysis of Consumption, Lifestyle, and Risk*, ed. R. Bunton, S. Nettleton, and R. Burrows (London: Routledge, 1995).
15. C. Shilling, *The Body and Social Theory* (London: Sage, 1993).
16. Maria D. Vesperi, "Introduction: Media, Marketing, and Images of the Older Person in the Information Age," *Special Issue: Images of Aging in Media and Marketing, Generations*, Fall 2001, 5–9; see also essays by Jacob J. Climo, "Images of Aging in Virtual Reality: The Internet and the Community of Affect," and Lee Bird Leavengood, "Older People and Internet Use" in the same issue.
17. See Anne Marie Balsamo, *Technologies of the Gendered Body: Reading Cyborg Women* (Durham, NC: Duke University Press, 1996).
18. Among the scholars who have written about the gender coding of domestic appliances is Ann Gray, *Video Playtime: The Gendering of a Leisure Technology* (London: Routledge, 1992).

19. See Jon F. Nussbaum, Loretta L. Pecchioni, James D. Robinson, and Teresa L. Thompson, eds., *Communication and Aging*, 2d ed. (Mahwah, N.J.: Lawrence Erlbaum Associates, 2000).
20. Marsha Cassidy, "Cyberspace meets Domestic Space: Personal Computers, women's work, and the Gendered Territories of the Family Home," *Critical Studies in Media Communication*, 18:1, 44–65.
21. *My Generation*'s editorial direction would mirror the Baby Boomers' stance toward old age: continue to defer it into some imagined future. *My Generation* began with 50–55-year-old AARP members as its audience in 2001. The plan was to add 56-year-olds in 2002, 57-year-olds in 2003, until the time would come when *Modern Maturity* simply would, well, expire. See David Rakoff, *Brandweek*, 42, no. 10 (March 5, 2001): 18–19.
22. Lorraine Calvacca, "Magazines Grow with the Market," *Advertising Age*, 17, no. 29 (July 10, 2000): 1–2.
23. David Goetzl, "Beltone Sees Big Market in Graying Baby Boomer," *Advertising Age*, 70, no. 15 (April 5, 1999): 3–4.

Chapter 8

1. Ken Dychtwald, *AgePower: How the 21st Century Will Be Ruled by the New Old* (New York: Jeremy P. Tarcher/Putnam, 1999), 110. Emphasis added.
2. Among the books on aging that became bestsellers in the 1990s were: Betty Freidan, *The Fountain of Age* (New York: Simon and Schuster, 1993); Tracy Kidder, *Old Friends* (New York: Houghton Mifflin, 1993); Hugh Downs, *Fifty to Forever* (Nashville, Tenn.: Thomas Nelson Publishers, 1994); Gail Sheehy, *New Passages: Mapping Your Life Across Time* (New York: HarperCollins, 1995); Studs Terkel, *Coming of Age: The Story of Our Century By Those Who've Lived It* (New York: St. Martin's Griffin, 1995); Jimmy Carter, *The Virtues of Aging* (New York: Ballantine Books, 1998); and Mary Pipher, *Another Country: Navigating the Emotional Terrain of Our Elders* (New York: Riverhead Books, 1999).
3. Matilda White Riley, Robert L. Kahn, and Anne Foner, eds., *Age and Structural Lag: Society's Failure to Provide Meaningful Opportunities in Work, Family, and Leisure*, (New York: John Wiley & Sons, 1994).
4. Douglas Coupland, *Generation X: Tales for an Accelerated Culture* (New York: St. Martin's Press, 1992).
5. William H. Whyte's *The Organization Man*, first published in 1956, was rereleased in 2002 (Philadelphia: University of Pennsylvania Press). The book has been heralded as an important explanation of the generation of American workers whose conformity to organizational life both fueled American business dominance and suppressed individuality and purposefulness among the workforce. The "organization man" of Whyte's day is easily read as today's traditionalists.
6. Donna Haraway, for one, takes up the issue of "home-work" as a result of the feminized restructuring of work, subjecting employees "to time arrangements on and off the paid job that make mockery of a limited work day," in *Simians, Cyborgs, and Women: The Reinvention of Nature* (New York: Routledge, 1991), 166. Commenting on the new permeability of workplace/home boundaries in the information economy, Haraway stresses that this condition is made possible by, but not caused by, the new technologies themselves. Instead, patriarchal capitalism's hold on the worker's postmodern identity is a major cause.
7. See Arlie Russell Hochschild, *The Time Bind: When Work Becomes Home and Home Becomes Work* (New York: Metropolitan Books, 1997).
8. Myra Dinnerstein, *Women Between Two Worlds: Midlife Reflections on Work and Family* (Philadelphia: Temple University Press, 1992).
9. Juliet Schor, *The Overworked American: The Unexpected Decline of Leisure* (New York: Basic Books, 1992).
10. Martin Sicker, *The Political Economy of Work in the 21st Century: Implications for an Aging American Workforce*, Westport, Conn.: Quorum Books, 2002. Sicker's conclusions are drawn from Dun and Bradstreet data.
11. Ibid., 37.
12. Ibid.; and William J. Baumol and Edward N. Wolff, *Protracted Frictional Unemployment as a Heavy Cost of Technical Progress*, Economics Working Paper Archive at WUSTL, March 1998.

13. John Hartley, in *Tele-ology* (London: Routledge, 1992), introduces the idea that television discourse is "paedocratic:" It positions the audience (the dominant element of which is Baby Boomers) as children, whose subjectivity must be regulated by the discursive adult. Similarly, with the management discourse on Boomers, supervisors (positioned as adults, even if they are Boomers themselves), are taught how to regulate the behavior of these "eternal children" for maximum workplace efficacy.

Chapter 9

1. Pamela J. Kalbfleisch, *Mentoring as a Personal Relationship* (New York: Guilford, in press).
2. Pamela J. Kalbfleisch and Arlyn Anderson, "Mentoring Across Generations: Culture, Family and Mentoring Relationships," in *Cross-Cultural Communication and Aging in the United States*, ed. Hana S. Noor Al-Deen (Mahwah, N.J.: Lawrence Erlbaum Associates, 1997), 97–120.
3. Richard Wiscott and Karen Kopera-Frye, "Sharing of Culture: Adult Grandchildren's Perceptions of Intergenerational Relations," *International Journal of Aging and Human Development*, 51, no. 3: 199–215.
4. Angie Williams and Jon F. Nussbaum have addressed the potential awkwardness involved in reconstructing adult child-parent relationships as circumstances change, because, unlike intergenerational workplace relationships, these family interactions are encrusted with complicated emotional histories. See Williams and Nussbaum, *Intergenerational Communication Across the Life Span* (Mahwah, N.J.: Lawrence Erlbaum Associates, 2001).
5. See V.L. Bengston and R.A. Harootyan, *Intergenerational Linkages: Hidden Connections in American Society* (New York: Springer, 1994); and B. Cohler, "Autonomy and Interdependence in the Family of Adulthood," *The Gerontologist*, 23 (1983): 333–39; and T.K. Hareven, *Aging and Generational Relations: Life-Course and Cross-Cultural Perspectives* (New York: Aldine De Gruyter, 1996).
6. Sara J. Czaja and Chin Chin Lee, "The Internet and Older Adults: Design Challenges and Opportunities," in *Communication, Technology and Aging: Opportunities and Challenges for the Future*, ed. Neil Charness, Denise C. Park, and Bernhard A. Sabel (New York: Springer Publishing, 2001).
7. Martin Kohli, "Work and Retirement: A Comparative Perspective," in *Age and Structural Lag: Societies Failure to Provide Meaningful Opportunities for Work, Family, and Leisure*, ed. Matilda White Riley, Robert L. Kahn, and Anne Foner (New York: John Wiley & Sons, 1994).
8. Stephen Katz, *Disciplining Old Age: The Formation of Gerontological Knowledge* (Charlottesville: University of Virginia Press, 1996); and P.C. Kontas, "Resisting Institutionalization: Constructing Old Age and Negotiating Home," *Journal of Aging Studies*, 12, no. 2: 167–84, and cited in Emmanuelle Tulle-Winton, "Old Bodies," in *The Body, Culture and Society: An Introduction* ed. Philip Hancock et al., (Buckingham, UK: Open University Press, 2000).
9. Tulle-Winton, "Old Bodies." p. 82.
10. Reinhold Kliegl, Doris Philipp, Matthias Luckner, and Ralf Th. Krampe, "Face Memory Skill Acquisition," in *Communication, Technology and Aging: Opportunities and Challenges for the Future*, ed. Neil Charness et al., 169–86.
11. R.W. Morrell, C.B. Mayhorn, and J. Bennett, "A Survey of World Wide Web Use in Middle-aged and Older Adults," *Human Factors*, in press; and S.J. Czaja, "Aging and the Acquisition of Computer Skills," in *Aging and Skilled Performance: Advances in Theory and Applications*, ed. W.A. Rogers, A.D. Fisk, and N. Walker (Mahwah, N.J.: Lawrence Erlbaum Associates), 241–66.
12. Wendy A. Rogers, Regan H. Campbell, and Richard Pak, "A Systems Approach for Training Older Adults to Use Technology," in *Communication, Technology and Aging: Opportunities and Challenges for the Future*, ed. Neil Charness et al., 187–208.
13. Hans-Werner Wahl and Clemens Tesch-Romer, "Aging, Sensory Loss, and Social Functioning," in ibid., 108–26.
14. Ronald P. Mannheim, "Generations Learning Together," in *Intergenerational Approaches in Aging: Implications for Education, Policy and Practice*, ed. Kevin Brabazon and Robert Disch (New York: Haworth Press, 1997), 79–91.

Chapter 10

1. International Longevity Center–US, *Aging on the World Stage,* April 5–8, 2002, Madrid, summary of conference proceedings.
2. "Falling Through the Net: a Report on Americans access to Technology Tools," 2001, *www.ntia.doc.gov/ntiahome/dn/*
3. *www.pewinternet.org,* "Pew Internet and American Life."
4. *www.commerce.net/research/stats/wwstats.html*
5. "A Nation Online: How Americans are Expanding Their Use of the Internet," 2002, *www.ntia.doc.gov/ntiahome/dn/*
6. "Falling Through the Net: a Report on Americans access to Technology Tools," 2001, *www.ntia.doc.gov/ntiahome/dn/*
7. "A Nation Online."
8. Ibid.
9. Ibid.
10. See data in Eric C. Newburger, "Home Computers and Internet Use in the United States, August 2000," *Current Population Reports,* U.S. Bureau of the Census, 2001. *www.census.gov.*
11. "Falling Through the Net," "A Nation Online."
12. Anne Cooper-Chen, personal communication, May 2002.
13. International Longevity Center–US, *Aging on the World Stage.*
14. Kevin Kinsella and Victoria A. Velkoff, "An Aging World: 2001," U.S. Census. *Bureau International Population Reports,* November 2001. *www.census.gov.*
15. Ibid.
16. Ibid., p. 3.
17. Gerard Goggin and Christopher Newell, *Digital Disability: The Social Construction of Disability in New Media* (Lanham, MD: Rowan and Littlefield, 2003).
18. "Falling Through the Net."
19. Ibid.
20. Age Concern England, *The Debate of the Age: Summary of Participation* (London: author, 2000), 81.
21. International Longevity Center, USA, *www.ilc.org.*
22. Kinsella and Velkoff, "An Aging World."
23. Alexander Samorodov, "Employment and Training Papers 33: Ageing and Labour markets for older workers," International Labour Organisation, 1999, *www.ilo.org/public/English/employment/strat/pub/etp33.htm*
24. Martha N. Ozawa and Huann-yui Tseng, "Differences in Net Worth Between Elderly Black People and Elderly White People," *Social Work Research,* 24, no. 2.
25. U.S. Department of Labuor, "Report on the American Workforce 2001," *http://www.bls.gov/opub/rtaw/rtawhome.htm,* accessed July 2003.
26. Kinsella and Velkoff, "An Aging World."
27. Ibid.
28. International Longevity Center-USA, *www.ilc.org.*
29. Ibid.
30. Rand, "U.S. Department of Health and Human Services, Administration on Aging, A Profile of Older Americans: 2001," *www.rand.org/publications/RB/RB5046,* January 2003.
31. Ibid. Americans age 65 and over have household computers and use the Internet at slightly less than half the number of Americans generally. In 2001, about half of American households had home computers, but only 24 percent of households headed by those 65 and over had home computers.
32. Ibid.
33. Ibid.
34. *www.iwpr.org/research_poverty.html*
35. For example, see Robert Butler and Herbert Gleason, eds., *Productive Aging* (New York: Springer Verlag, 1985).
36. Samorodov, "Employment and Training Papers."
37. Age Concern England, *Debate of the Age: Future Work and Lifestyles* (London: author, 2001).
38. *www.agepositive.gov.uk* and *www.newdeal.gov.uk*

39. Ginn and S. Arber, 1996. "Patterns of Employment, Gender, and Pensions: The Effect of Work History on Older Women's Non-state Pensions," *Work, Employment and Society*, 10, no. 3: 469–90.
40. Colette V. Browne, *Women, Feminism and Aging* (New York: Springer Publishing Co., 1998).
41. Age Concern England, *Debate of the Age: Future Work and Lifestyles*.
42. Ibid.
43. Ibid.
44. See Bland, Carole J. and Berquist, William H., "The Vitality of Senior Faculty Members," 1997, *Eric Digests*, http://www.ericfacility.net/databases/ERIC-Digests/ed 415733.html.
45. See "Metropolitan Universities: An International Forum," http://muj.uc.iupui.edu/7_4.asp, accessed Aug. 29, 2003.
46. Ernest L. Boyer, *Scholarship Reconsidered: Priorities of the Professorate* (Princeton, N.J.: Carnegie Foundation for the Advancement of Learning, 1990). See examples of resulting work in William G. Tierney, ed., *The Responsive University: Restructuring for High Performance* (Baltimore: Johns Hopkins University Press, 1998); and Barbara Leigh Smmith and John McCann, eds., *Reinventing Ourselves: Interdisciplinary Education, Collaborative Learning, and Experimentation in Higher Education* (Boston, Mass.: Anker Publishing, 2001).
47. "Age Discrimination: A Pervasive and Damaging Influence," U.S. Administration on Aging, *www.aoa.gov*, viewed January 10, 2003; and "Age Discrimination in Employment Act (ADEA) Charges, FY 1992-FY 2001, *www.eeoc.gov*, viewed January 10, 2003.
48. Samorodov, "Employment and Training Papers."
49. Ibid.
50. Ibid.
51. Ibid.
52. Ibid.
53. Ibid.
54. Lisa Nakamura, *Cybertypes: Race, Ethnicity, and Identity on the Internet* (New York: Routledge, 2002).
55. Cathy Horyn and Suzanne Kapner, "Burberry Finds a Fountain of Youth," *New York Times*, July 12, 2002, W1, 3.
56. Goggin and Newell, *Digital Disability*, xv.
57. UK Department of Trade and Industry, *Ageing Population Panel, 5: Applications of Information and Communications Technology Taskforce*, 2000, 2.
58. Ibid.
59. See, for example, Peter Senge et al., *The Dance of Change: The Challenges to Sustaining Momentum in Learning Organizations* (New York: Currency/Doubleday, 1999).
60. For a range of examples, see Smith and McCann, eds., *Reinventing Ourselves*.
61. *The Social Policy of Old Age: Moving into the 21st Century*, ed. Miriam Bernard and Judith Phillips (London: Centre of Policy on Ageing, 1998).
62. Miriam Bernard and Judith Phillips, "The Challenge of Ageing in Tomorrow's Britain," *Ageing and Society*, 20 (2000): 33–54.
63. Meredith Minkler, "Critical Perspectives on Ageing: New Challenges for Gerontology," *Ageing and Society*, 16, no. 4: 470.
64. Maria Evandrou, "Great Expectations: Social Policy and the New Millennium Elders," in Bernard and Phillips, eds., *The Social Policy of Old Age*; and Bernard and Phillips, "The Challenge of Ageing in Tomorrow's Britain." 33–54.
65. Margaret Clark, "The Anthropology of Aging, a New Area for Studies of Culture and Personality," *Gerontologist* 7 (1967): 55–64.
66. Linda Sax, Alexander Astin, William Korn, and Shannon Gilmartin, *The American College Teacher* (Los Angeles: Higher Education Research Institute of the University of California at Los Angeles, 1999).

Afterword

More than anything, this book has been about activism, both in the sense that I have meant to recognize the involvement on the part of and on behalf of aging Americans and in the sense that I am proposing that we see more such work. I have described the experiences of aging workers with respect to technology, and I have suggested changes for what seem to me to represent the promise of better work environments for people of different generations who now find themselves in teams together or reporting across unanticipated age lines.

It would seem most appropriate if some of these arguments were taken up by graduate students, who will carve out the "next generation" of applied communication and business models in our technologically driven society. Students who begin to think about age/work issues today in the interest of presenting tomorrow's solutions will have an enthusiastic following. Students in broader areas of gerontology who begin to think of linking work with age in their own thinking, rather than only the aging-retirement nexus, will find fertile ground to plow in their own disciplines.

As with my previous work, I am hopeful that, through this book, readers will find ways to encourage more positive understandings of relations between the generations. It is well past time for young people to see the body of an older adult for what it is—the weathering of a life—than as a rotting shell. The many people I had the fortune to encounter for the purpose of preparing this book displayed rich experiences, idiomatic wisdom, and a purposefulness toward living.

The thing that has struck me most in my encounters with the elders I met in connection with this book is the tension so many of them have expressed to me between their passion for achievement and the blows that their dignity has suffered following setbacks from the digital economy. Many, as the chapters here reveal, have emerged successfully from their challenges. Others have been less fortunate, sometimes due to cultural constraints and sometimes, I am tempted to think, more due to a lack of self-awareness with regard to their own personal characteristics and abilities.

As suggested in Chapter 5, all of us must work harder to help one another grow as "self-programmable elders." The time for resting on past accomplishments is past. And the time for reacting to the well-identified "megatrend" is, as the kids used to say, so fifteen minutes ago. Now is the time for us to constantly monitor our social environments and brainstorm with diverse others to make sense of our strategic planning and everyday tactics. We know little of tomorrow's workplaces other than that the people we will find there will represent diverse interests, including generations. It will be much better for us to complement one another's differences than to compete against others for scarce resources.

In May 2003, the Bush administration changed the formula for calculating the monetary value of a human life for the purposes of public spending on health. For example, when the Environmental Protection Agency would calculate where to place scarce resources for public health, it would turn to demographic tables to determine the dollar benefit of lives at risk or lives to be saved. The idea of this change was that old people's lives are not as valuable as young people's, considering that they are so much closer to death. A cry of protests quickly rang out from elder advocates, who expressed anger over the so-called senior death discount. The Bush administration quickly backed off, but the conclusion is clear: Institutions often engage in pitting the generations against one another through the fueling of self-interest. To state that a 60-year-old's life was less valuable than a 30-year-old's hearkens to the time when 60-year-olds might have said, "I have lived my life. Now it's your turn." I'm not sure that today's 60-year-olds would agree with such sentiment. Life is valuable, no matter whose.

Index

AARP (American Association of Retired Persons): 5, 48, 158, 177, 216, 218, 225; criticized for business interests, 47
aarp.org: 22–7, 47, 83, 208, 214; lobbying on Medicare and Social Security benefits, 48; Money and Work message board, 29, 39–41; "Working Options: A Guide for Midlife and Older Workers," 23–4; Working Options message board, 43–4, 51
About Schmidt, 118, 154
advertising: elder inclusive, 81–2; technology, 157
African American: female interview subjects' deference to white interviewer, 98
African American studies: as bookstore section, 180
Age Advantage, The, 205
Age Concern England, 229, 232–3, 235
age discrimination: 7, 38, 89–90; discussed on aarp.org, 23; in downsizing/layoffs, 42–5; encouraging others to sue, 44; fear of, 28; governments ineffective against, 236–7; relatively young worry over, 46; as root of most older workers' problems, 31; venting on message boards, 42–3; victim status, 45
Age Discrimination in Employment Act, 7, 24, 194, 236
ageism, 7, 11, 37, 83–4, 109–10, 112
age-gender intersection, 196
AgePower, 179
Agequake, 205
age-race-ethnicity-class intersection, 149
age-work-technology intersection: 10, 14, 119, 215, 226, 242; in books, 180; enhancement in the United States, 232–41; in films, 124, 130, 142; in magazines, 164–78; 196
aging: 4–9; books on, 4, 179–81, 198–9, 205–6; difficulty of portraying in American culture, 12; researching,

240; technology workers, 238; workforce, 182, 225
aging/work message boards, 32–55
AIDS, 84, 95
Altavista: career advice, 27
Amazon.com: 136, 181; books on, 4
Ameche, Don, 127
America the Wise, 205
American Demographics, 82
Americans with Disabilities Act, 194
Anderson, Arlyn, 219
Andrus Gerontology Center, 114
Aniston, Jennifer, 161
Another Country, 179
Apple, 111
Arthritis, 8

Baby Boom: 196, 228; changing perception of retirement for the generation, 2; largest generation, 196
Baby Boomers: 3, 4, 14, 31–2, 99–100, 105, 112–13, 178–9, 183, 218, 230, 237, 241–2; aging of, 5, 23, 155, 227; coming of age, 130; in contrast to joystick generation, 103–4; in Hollywood, 118; identity exploration, 154; late in career, 158; management of, 182, 194–5; maturity, 205; out of work in soft economy, 21–2; as parents, 193, 197; as posters on ThirdAge.com message board, 41; questioning authority, 47; readers of *Rolling Stone*, 166; and retirement, 177, 199; and technology, 194–5; women, 161–2, 196–7
Barnes and Noble, 181
Back to the Future, 137–8, 142
Basinger, Kim, 137
Baumol, William J., 198
Beauvoir, Simone de, 57
Benjamin, Walter, 119
Bennis, Warren G., 192
Bernard, Miriam, 240
Better Homes and Gardens, 178
Black Rocket, 164–5

Blaikie, Andrew, 117
Blondell, Joan, 147
Bn.com, 181
Bond, James, 117–18, 130, 136
Bonet, Lisa, 151
Botox Cosmetic, 161–2
Boyer, Ernest, 236
Brand, Dionne, 83
Brandauer, Klaus Maria, 136
Bravo, Rose Marie, 238
Brimley, Wilford, 127
Brokaw, Tom, 4
Bryson, Dee, 39
Buchanan, Pat, 34
Buffy the Vampire Slayer, 12
Burberry, 238
Burgess, Emma, 205
Bush, George W., 46, 258
Business Week, 165
Butler, Robert, 7–8

Cadillac Esplanade, 178
Canada, 236
Canja, Esther, 225
capitalism: costs and benefits, 34; and patriarchy, 65; social realities of, 86
Careerpath.com: career advice, 27–8
Carlsson, Chris, 9
Carrera, Barbara, 136
Carter, Jimmy, 4
Cassidy, Marsha, 176
Castells, Manuel, 104–6, 113–15
Center for Twentieth Century Studies, 14
Center for Twenty-First Century Studies, 14
Charisse, Cyd, 118
Charles Schwab, 167
Charlie's Angels, 154
Charter of Rights and Freedoms: of Canada, 7
Cinderella, 125–26
Clark, Margaret, 13, 240
Class, 10, 12–13, 17–18, 73, 83, 107, 173, 229; and Digital Divide, 226; as dividing structure, 242
Climo, Jacob, 175
Clinton, Bill, 46
Clurman, Ann, 199
CNN, 208
Coastal Living, 173
Cockburn, Cynthia, 10, 67
Cocoon, 18, 117, 127–28, 165
Coleman, Dabney, 137

colonialism: social realities of, 86
Coltrane, Robbie, 129
Coming of Age, 179
commodity spiritualism, 41
Congress: 35; unemployment benefits debate, 49
Connery, Sean, 117–18, 130, 136, 138
Consalvo, Mia, 33
Contact, 153–4
Conversation, The, 147–8
Coppola, Francis Ford, 147–8
Corbin, Barry, 137
Coupland, Douglas, 183
Cromwell, James, 151
Cronyn, Hume, 127
Crouching Tiger: Hidden Dragon, 153
Cruise, Tom, 139, 155
C-SPAN, 4
cultural research, 13
cyberia, 103, 109
Czaja, Sara J., 220

Davidson, Arthur, 60
Davis, Richard, 28
Death Becomes Her, 18, 128–9
debit cards, 4
Dell, Michael, 194
dementia, 8, 82
demographics, 157, 181–2, 193, 225, 230
De Niro, Robert, 154
Dennehy, Brian, 128
Dennis, Helen, 114
depression, 8, 53, 198
Desk Set, The, 142, 147
Diamonds Are Forever, 136
Digital Divide, 10, 12, 207, 225, 227–8, 231; contributing factors, 226
Digital Work Force, The, 11
Direct Marketing, 81–2
disabilities: 195; and labor force, 230; as message board topic, 40; as obstacle to Internet, 229
disability benefit programs, 3, 231
disabled: 238–9; elder depicted in film, 141
Disney, Richard, 6
divorce, 198
Do, Thuy-Phuong, 15
Downs, Hugh, 179
Driving Miss Daisy, 117, 148–9, 154–5
Drucker, Peter, 182
drug abuse: 84, 95
Duening, Thomas N., 192

During, Simon, 119
Dychtwald, Ken, 179
Dylan, Bob, 32

Eastwood, Clint, 118, 151, 154
Easy Rider, 60
economic inequality: among the elderly, 6
education: 235; and Digital Divide, 226
education, higher, 235–6, 239; scholarship on technology, 240
email, 4, 25, 30, 43, 59, 62–3, 68, 70, 158, 162–3, 176, 181, 193, 197, 219
employees: use of temporaries, 58
Enemy of the State, 150
Entertainment Weekly, 166
Environmental Protection Agency, 258
Equal Employment Opportunity Commission, 24, 26, 236
ergonomics, 195, 220
Estes, Carroll L., 8
Ethnicity, 10, 12, 28, 173, 237; and Digital Divide, 226; as dividing structure, 242; importance in Harley-Davidson social organization, 79
European Federation of the Elderly, 228

fairy tales: film theme, 118–30, 153
Featherstone, Mike, 32, 174
feminism: Second Wave, 58
feminist theories: and women at Harley-Davidson, 67
fertility rates, lowering, 4, 228
Fifty to Forever, 179
Fifty-plus and Looking for Love Online, 205
Filipczak, Bob, 192
films: and aging bodies, 118; targeted to young and typically male audiences, 117; treatment of again, work, and technology, 124
Finland: policies on aging and employment, 233
flaming: on Monster.com's Age Issues message board, 32–4; over self-pity, 45
Ford, Harrison, 138, 148
Fortune, 173
Foster, Jodie, 153
Foucault, Michel, 107–8, 173
Fountain of Age, The, 179
Fourth Turning, The, 205
Fox, Michael J., 137
Fraser, Jill Andresky, 22
Freeman, Morgan, 130, 148

Fried Green Tomatoes, 117
Friedan, Betty, 4, 179
Future of Work, The, 192

Gameboy Advance, 165
Garner, James, 151
Garr, Teri, 148
Gates, Bill, 9, 100, 106, 194
Geeks and Geezers, 192
Geist, Patricia, 15
gender: 10, 12–13, 18, 28, 107, 173, 229, 237; behavioral guidelines, 48; biased workplaces, 83; complaints on Monster.com's message board about bias, 33; consciousness, 86; differences at work, 66; differences in accepting computers, 66; as dividing structure, 242; informing roles of elders in fairy tales, 119; software selection, 69; technological innovation and existing relations, 67–8; value of feminine appearance in workplace, 73
gender studies: as bookstore section, 180
Generation X: 12, 15, 17, 51, 157, 166, 209, 219, 226, 242; explaining its members, 192–3; geeks, 150; and Harley-Davidson clothing, 59; interplay with other generations in workplace, 182–3, 194, 197; in magazine advertising, 164; sharing Internet with elders, 208–9, 211–23; and technological competency of, 183, 195; work expectations, 106
Generation X, 183
Generation Y: 15, 17, 161, 196; and work expectations, 106
generations, 193
Generations Apart, 205
Generations at Work, 192
generic labor, 107–8, 112–14
gerontology, 220, 240, 257
"girl power," 12
Gitlin, Todd, 207–8
Glenn, John, 4, 152
Goddesses in Older Women, 205
Goggin, Gerard, 238
Goffman, Erving, 29, 99
grandparents: raising grandchildren, 84
Grant, Lou, 32
Grantham, Charles, 192
Gray Dawn, 205
"gray market," 238

Gray Panthers, 240
Grosz, Elizabeth, 171
Guinness, Alec, 126

Hackman, Gene, 148, 150
Hamill, Mark, 126
Haraway, Donna, 65, 86, 97
Harley, Bill, 60
Harley-Davidson, 59–80; complex social organization, 78; design process, 63; female employees' use of computers, 63–80; female employees' repeated adjustment to new software, 69–70; as symbols of male prowess and freedom to roam, 60; Talladega Test Facility, 59; women taking work home, 74–5; "women's work" often involves transfers, 65
Harley Owners Group, 60
Harris, Richard, 129
Harris pagination system, 111
Harry Potter and the Sorcerer's Stone, 129–30
Hawn, Goldie, 128
health plan: employers' cost for, 6
Hebdige, Dick, 175
Heflin, Jay S., 205
Hepburn, Katharine, 142, 147
Hewlett-Packard, 162
Holmlund, Chris, 170
home telecommuting, 234
hooks, bell, 12
Howe, Neil, 205
Hurt, John, 153
Hutton, Lauren, 161

Images of Aging in Media and Marketing, 175
immigration: blamed for job losses, 46
incarceration: 84, 95
income: and Digital Divide, 226
Indiana Jones and the Last Crusade, 138–9
Individual Retirement Accounts (IRAs), 3
Information Revolution, 4
Information Superhighway: 15; and aging advocates, 22
information technology (IT): 229; labor market, 22; laid off from jobs in, 35; youth of managers in, 11
insurance: increased cost of, 6
interdependence: of society, 8
intergenerational communication: and the Internet, 207–9, 211–223; in the workplace, 28, 70, 76–7, 179–206

intergenerational friction: in the workplace, 51
intergenerational learning communities, 239–40
intergenerational relationships, 14; in films, 130, 136–42, 152–3, 240, 257
intergenerational studies, 221
intergenerational workforce: 181–98; benefits of elders, 232; and smaller companies, 197–8
International Labour Organisation, 228; Older Workers Recommendation, 237
International Longevity Center—USA, 230
Internet: access 226; access at work, 74, 225; acquiring skills for use of , 9; aging/work message boards, 21, 29–55; career advice on, 23–9, 49, 52; culture of, 29; fear of, 88, 95–6; as gunpowder, 10; intergenerational sharing of, 208–9, 211–23; marketing to over-55s, 83; notion of community, 50; older users, 222–3; as an opponent, 14; query for stories, 18; research results from, 16; starting a new business through, 54; study of use, 19; time spent on, 1; use of, 3, 15, 23; use of in Asia-Pacific region, 227; use of in Europe, 226–7; use of in Scandinavia, 227; use of in United States, 227; vernacular of, 16; use in work at home, 197; women make up majority of users, 83. See also World Wide Web, 12
In the Line of Fire, 153
Iomega, 169–70
Ireland: and aging workers, 232; policies on aging and employment, 233
Italy: declining fertility rates in, 4
Ivancevich, John M., 192

Jagger, Mick, 5
Japan: forced retirement in, 11
JavaScript, 194
Jones, James Earl, 118
Jones, Tommy Lee, 151
Joy Luck Club, The, 153
joystick generation, 103

Kachgal, Tara, 41
Kalbfleisch, Pamela, 219
Keigher, Sharon, 14
King, Martin Luther, Jr., 149
Kinsella, Kevin, 228

Kirby, Marjorie, 30
Kitt, Eartha, 158
Koch, Ed, 158
Kopera-Frye, Karen, 219
Kuhn, Thomas, 106

Lancaster, Lynne C., 192
Lane, Kara, 49
Lauters, Amy, 81, 98
Leavengood, Lee, 175
Leder, Drew, 15
Lee, Chin Chin, 220
leisure workers, 175
Leno, Jay, 152
liberal capitalism, 9
liberal individualism, 8
life expectancy: as of 1900, 4; increase in, 2, 4, 104
Lindbergh, Charlie 4
Lloyd, Christopher, 137
loneliness, 8
longevity, 198–9, 228
Longevity Revolution, 205
L'Oreal, 162
Lovelock, Peter, 40

McCartney, Paul, 5
McKellan, Ian, 140
McMullin, Julie, 7
McPherson, Tara, 54
Macwarehouse.com, 167
magazine advertising: images of technology in use, 157, 170–1; and older workers, 157–8, 237; older women rarely shown in workplace, 171; technology marketed across gender lines, 175–76
Managing Einsteins, 192
Managing Generation X, 192
Mannheimer, Ronald, 223
Markus, Katrin, 228
Marshall, Penny, 154
Marshall, Victor, 7
Martin, Melissa, 28
Masterpiece Theatre, 4
Mature Outlook, 177
Matures, 31–2, 105, 113, 177–9, 182–3, 194, 196–7, 242; in contrast to joystick generation, 103–4; and self-programmable labor, 106
MCI, 237
Me Generation, 5

Medicare: 5, 9, 82, 99; AARP's lobbying on, 48
media streams, 33
Meet the Parents, 154
Mellencamp, Pat, 119, 126
men: aging in the workplace, 57–8
Men in Black, 166
menopause, 198
mentoring relationships, 219–20
Meredith Publications, 177
Merleau-Ponty, Maurice, 15
message boards: community building, 53
Microsoft, 35, 47, 237; Excel, 88; Office, 98; Word, 88
middle age: changing definition of, 13
Mieszkowski, Katharine, 22
millenniums, 242
millennials, 196
Milwaukee Public School District, 91–2
Milwaukee, Wisconsin: 210, 227; computer training classes in, 82, 84–5; women in, 82, 84–99
Mission Impossible, 139, 142, 151
Mitchall, David, 100
Modern Maturity, 5, 158, 162–3, 174–75, 177, 216
Monster.com: 49; Age Issues message board, 29–30, 32–41, 45, 50–1, 53–4; criticized for business interests, 47
Montreal: study of garment workers in, 7
Moody, Harry R., 6
Moore, Roger, 130
More, 177
MSN.com: Aging Horizons message board, 29–30, 37, 39, 42, 45–7, 49
My Generation: magazine, 177; as magazine title, 5

Nader, Ralph, 34
Nakamura, Lisa, 237
national identity, 10
National Institute on Aging, 212
National Telecommunications and Information Administration, 225
Neutrogena Visibly Younger, 161
Nextel, 165
Next50.com, 162
Never Say Never Again, 117, 130, 136–7
Net Generation, 219
New Deal: in Britain, 233
New Choices, 177
New Passages, 179

New Pioneers, The, 183
new technologies, 9–11
New York Times, 238
New York Times Magazine, 158
Newman, Paul, 118
Newsweek, 178
Nicholson, Jack, 118, 154
Nike: use of commodity spiritualism, 41
Newell, Christopher, 238
Norris, Pippa, 9
Nutrifusion, 162

Octopussy, 130
Olay, 162
Older Americans Act: of 1965, 11, 17
older people: as consumers, 157; failure to adapt to computers, 216–17, 220; film images, 117–55; and new technology, 208
older workers: aging men in workforce, 230; depictions in magazine advertising, 158; discomfort with new technology, 168; failure to adapt to computers, 78, 97–8; 111–12; images of, 237; management of, 195; mastering new technology, 136; and technology, 196; women, 86–95
Oromwell, James, 149
Outlaws: motorcycle group, 60
outsourcing: of jobs, 58
Oxygen, 12

Paine, Thomas, 37
Palm Pilot, 57
Pampers, 178
patriarchy: and capitalism, 65; in film,147; social realities of, 86; and workplace, 83
PBS (Public Broadcasting System), 4
Peterson, Peter G., 205
Petzinger, Thomas, Jr., 183
Phenomenology, 15; of aging, 174
Phillips, Judith, 240
Phoenix Wealth Management, 165
Pierres, Susan, 158
Pipher, Mary, 179
PlayStation, 206
Playboy, 178
Policy, 225–42
Poster, Jamie, 30
Postman, Neil, 99
poverty: 8; feminization of, 7; feminization among elderly of, 7

PowerPoint, 15, 194
prejudice: 153; against elderly, 8; becoming a union steward to oppose, 87; silencing expressions of racial prejudice by becoming union steward, 68
Priceline.com, 23
"productive aging," 8, 234
productivity: among older workers, 6
Pumping Iron II, 170

Quark Xpress, 111

race: 10, 12–13, 18, 83–4, 107, 173, 193; consciousness, 86; and Digital Divide, 226; as dividing structure, 242; importance in Harley-Davidson social organization, 79; rarely mentioned on Monster.com's message board, 33
racism: challenging it in Milwaukee, 91–2
Raines, Claire, 192
Rand organization, 230
Red Hot Mamas, 205
Redgrave, Vanessa, 139, 151
Reeboks, 5
religion: 10; rarely mentioned on Monster.com's message board, 33
reproduction: lengthening potential of, 104
Reservoir Dogs, 166
retirement: benefits, 9; income, 41; programs, 3; reconceptualization of, 2–4; and senior Web surfer, 22; systems, 232–4
Retirement: compulsory, 236; United Kingdom legislation protecting elders from mandatory age 60 retirement, 22
Retirement for Dummies, 205
revisionist history in film: 142–53
Riley, Matilda W., 17, 180
Rocking the Ages, 199
Rogers, Wendy, 222
Rolling Stone, 166
Rolling Stones, 5
Rossellini, Isabella, 128
Roszak, Theodore, 205

sabbaticals, 235
Salon.com, 22
Samorodov, Alexander, 237
"senior citizen": changing meaning of, 5
self-programmable labor, 103, 105–12, 114–15

Senge, Peter, 239
Seniornet.com, 12, 22, 83
Sex After Sixty, 205
sexual orientation, 10; studies [as bookstore section], 180
sexuality, 10, 13, 173, 193
Shaping Women's Work (Webster), 10
Sheehy, Gail, 179
Shields, Rob, 103, 115
Sicker, Martin, 198
Siebert, Muriel, 174
Simey, Margaret, 240
Simon, Carly, 32
Skerritt, Tom, 153
Slack, Roger, 50
Smith, J. Walker, 199
Smith, Maggie, 129
Smith, Will, 150
Snyder, Sharon, 100
Sobchack, Vivian, 128–9
Social Security: 9, 86, 97, 199, 208, 231; Act of, 1935 2; AARP lobbying on benefits, 48; Benefits, 2, 3; pensioners, 14; Trust Fund, 3
Sony, 169–70
Space Cowboys, 18, 151–52, 154–5, 238
Spielberg, Steven, 138
Sports Illustrated, 178
Star Trek: First Contact, 149–50, 153
Star Wars, 126–7, 153
Stewart, Patrick, 140, 149, 153
Stillman, David, 192
Stone, Allucquére Rosanne, 40
Strauss, William, 205
Streep, Meryl, 128–9, 155
Sunbelt, 3, 82
Sutherland, Donald, 151
Sword in the Stone, The, 124–25

Tandy, Jessica, 118, 148
Tapscott, Donald, 183
Target, 165–6
technology: vs. humans, 147; elder playing with, 166; as liberating force, 153; as magic, 118; older people's struggles with, 150; older workers' discomfort with new, 168; and policy making, 240; scholarship on, 240; as tool for evil, 148, 150
teen pregnancy, 84
Terkel, Studs, 179
Thau, Richard D., 205
The Who, 5

ThirdAge.com: 22, 29–30, 40–1, 49, 83; Job Hunting Over 50 message board, 48
Thomas, Robert J., 192
Thunderball, 136
Tiernan, Jennifer, 53
Time, 158, 161
Tracy, Spencer, 142, 147
Traditionalists, 196–7
Transcast, 163
Travelocity.com, 23
Tulgan, Bruce, 192
Tulle-Winton, Emmanuelle, 174, 221–2
Turner, Tina, 5

unions: progressive relationship with Harley-Davidson, 61–2; represented at Harley-Davidson meetings, job interviews, 62
United Kingdom: and aging workers, 229, 232, 235–6; Amazon Web site for, 136; legislation protecting elders from mandatory age 60 retirement, 22
United Nations, 228, 230
United Parcel Service, 168, 174–5
United Technologies Research Center, 166
University of Wisconsin-Milwaukee, 14, 207
Ure, John, 40
U.S. Army: too old to join, 38
U.S. Census Bureau, 228
U.S. Department of Commerce: 11; Information Technology Work Force Convocation, 22
USA Today, 111, 152

Valetta, Amber, 161
Van Dyke, Dick, 126
Velkoff, Victoria A., 228
Vesperi, Maria D., 175
veterans: as message board topic, 40
Vietnam: female veterans, 53
Vinson, Jacquelyn, 81, 87, 98
virtual community: as context bound, 31
Vogue, 161–2, 178
Voight, Jon, 130, 139, 150, 154

Walker, Jean Erickson, 205
Wall Street Journal, 12
Wallace, Paul, 205
Wal-Mart, 3
Wang, Vera, 161
War Games, 137, 142, 154
Washington: business in, 34; information from, 4

Web sites: 208; aging/work message boards, 21, 29–55; corporate, 198; health related, 82, 199, 211–13, 216–17; job related, 32; links to journalism and government sites, 35; of online booksellers, 181; National Institute on Aging, 22; personal, 30; for professional consulting services, 230; U.S. Administration on Aging, 22
Webster, Juliet, 10, 65, 79
WebTV, 162
Wernick, Andrew, 174
When Generations Collide, 192
Whole World Is Watching, The, 207
Wiener, Linda, 34
Williams, Roger A., 50
Willis, Bruce, 118
Wiscott, Richard, 219
Wolff, Edward N., 198

women: adapting to computerized workplace at midlife, 68; aging in the workplace, 58; in the workplace, 229; as Harley-Davidson employees, 59–80; work feminized through microelectronics, 65
Wood, John, 137
Woodward, Kathleen, 14, 172, 174
work 4–9; in elders' lives, 181; and technology, 194
World Wide Web, 12

X-Men, 139–42, 153, 155

You Know You're Getting Old When..., 206
Young Elders: aged 60–70, 82

Zemeckis, Robert, 153
Zemke, Ron, 192